A PERSPECTIVE ON OPIOID ADDICTION

A PERSPECTIVE ON OPIOID ADDICTION

JAY SCHULKIN
AND
BRYCE HUEBNER

Columbia University Press   *New York*

Columbia University Press
Publishers Since 1893
New York   Chichester, West Sussex

Library of Congress Cataloging-in-Publication Data
Names: Schulkin, Jay author | Huebner, Bryce author
Title: A perspective on opioid addiction / Jay Schulkin and
Bryce Huebner.
Description: New York : Columbia University Press, [2025] |
Includes bibliographical references and index.
Identifiers: LCCN 2025000682 | ISBN 9780231220644 hardback |
ISBN 9780231220651 trade paperback | ISBN 9780231563345 ebook
Subjects: LCSH: Opioid abuse—United States—History |
Opium—United States—History | Opioid abuse—Social aspects—
United States—History | Opioid abuse—Treatment—
United States—History | Opium—Physiological effect
Classification: LCC RC568.O45 S38 2025 |
DDC 362.29/3—dc23/eng/20250606

Cover design: Elliott S. Cairns
Cover image: Vitalii Vodolazskyi / Shutterstock.com

GPSR Authorized Representative: Easy Access System Europe,
Mustamäe tee 50, 10621 Tallinn, Estonia, gpsr.requests@easproject.com

CONTENTS

PREFACE

Across numerous historical and social contexts, people have experienced the aversive implications of opioid addiction. In its current guise, the opioid crisis often feels intractable. The opioids that are currently available are more deadly than those that have been previously available. Moreover, the scale of the current crisis, as well as its corporate facilitation, is absolutely breathtaking. By clarifying the socially entangled nature of opium use and abuse into view and by highlighting the complexity of the biology of addiction, we hope to bring better strategies for navigating the current crisis into view. But before we get started, it will help to say what we take addiction to be.

Addiction is *a brain-based phenomenon*. The search for opioids can dominate the activity of the amygdala, an almond-shaped structure tied to navigating various needs and challenges. At the same time, processes centered on the nucleus accumbens—a midbrain structure—can become tightly associated with drug use and reward. Furthermore, diverse processes ranging from the brainstem to the forebrain can be recruited to sustain the pursuit and use of opioids. The collective activity of these systems shapes action and attention. In the context of opioid addiction, the focus of these processes often becomes narrowed to yield an all-consuming drive to consume this substance. The result is a form of *cephalic dysregulation*, which affects processes throughout the body and pervasively shapes patterns of thought and action. Over time, however, the overactivation of these processes leads to neural decay and diminished control over bodily and cognitive processes. So, addiction should also be understood as *a deeply embodied phenomenon*. Just as importantly, addiction often yields a shrunken world where patterns of thought and action foster a vicious cycle of collapse. The all-consuming drive to consume opioids can make it difficult to keep a job, maintain loving relationships, or stay in school. It can lead some people to cheat, steal, or engage in other risky behaviors to obtain opioids. This can begin to close off opportunities for pursuing recovery and sobriety, it can make it difficult to take advantage of any opportunities there happen to be, and often, it can lead to a situation where others begin to reject, scold, and attack those who are struggling with addiction. Put bluntly, many forms of social stigma go hand in hand with addiction—so the manifestation of addiction is typically *social* and profoundly *existential*.

The interactions between the physiological, psychosocial, and social aspects of addiction often yield a pair of deep ironies. On the one hand, opioids can placate the scourge of pain, but attempts to manage pain can quickly slide into addiction—especially where modern opioids are used. On the other hand, the kinds of traumas that are commonly experienced by people in socially and economically disadvantaged situations can also be placated with opioids, but as a result, addiction to opioids—and overdoses from substances like fentanyl—become increasingly common among those who have the least available resources for navigating addiction successfully. This book will not solve these issues; but it might draw attention to the complexity of our current situation.

We want to apologize to any researchers whose work we have missed. The field is enormous, and we have only offered a perspective on a small part of it. This perspective on opioid addiction is neither complete nor exhaustive, but we hope it will be helpful. Jay cared deeply about this book. He and I discussed it for a long time, and he had already carried out much of the initial research when he learned that he had hepatic cancer. He sent me the draft that he had and asked me to help him finish it. So I quickly began to rewrite, reorganize, and reframe the arguments in ways that would pull his perspective on opioid addiction more clearly into view. Unfortunately, Jay died before I was able to finish the revisions. I have done my best to present the ideas in this book in a clear and accessible way. I can't be sure that I have succeeded in capturing Jay's aspirations, but this was my primary aim. This is not the book that I would have written on my own, but I hope it's a fair approximation of the book that Jay wanted to dedicate to his old friend from the Bronx, Billy Toth, who died over fifty years ago from an overdose of heroin.

Bryce Huebner
Washington, DC

A PERSPECTIVE ON OPIOID ADDICTION

INTRODUCTION

A Complex Story About Biology and Culture

Throughout our evolutionary history, humans have dedicated time and effort to discovering which plants are edible and which are useful for other purposes. This form of epistemic foraging has been a distinctive feature of human cognition. Across many different environments, we have identified beautiful flowers and determined which can serve medicinal or nutritional purposes. Humans have been very good at finding and using plants that reduce pain. This is no easy task. Over hundreds of millions of years of evolution, nature has produced an incredible array of psychoactive chemicals. Opium was originally derived from poppy plants, and long before it was synthesized, there was a complex interplay between nature and culture that shaped patterns of use and abuse. In part, this is because poppies are seasonal plants that require rich soil, sufficient sunlight, and sufficient water to grow (figure I.1). Both *P. somniferum* and *P. bracteatum* produce significant amounts of opium. But for much of human history, these plants were only available at certain times of the year and in specific regions.

At some point in our history, however, humans moved beyond recognizing the ameliorative properties of the poppy, and they started to cultivate this flower so that it could be used

FIGURE I.1 An opium poppy (*Papaver somniferum*).
Wikimedia Commons. Public domain image.

in a wider range of contexts at any point during the year. This is also a distinctive feature of our species. We collect, cultivate, and store plants, and we develop folk-biological systems that help us survive and flourish in the ecological contexts in which we live and act.[1] Perhaps most importantly, we have also shared this knowledge within broader epistemic communities, and this has supported biological and cultural evolution.

The story about our entanglement with the opium poppy is complex, and there is no way to tell this story in a single chapter or even a single book. Thus, our aim in this introduction is simply to introduce the idea that human capacities to explore, test, and catalog have had a significant impact on our evolved vulnerability to addiction. We argue that humans are distinctively vulnerable to opioid addiction because we share an evolutionary history that has made us both curious and able to cultivate rich patterns of cultural stability. Evolution is not a linear process. Diverse processes produce a range of possibilities, and those that consistently compromise the pursuit of viability in a specific ecological niche tend to be eliminated. However, just as importantly, humans are niche constructors who engineer their environments in ways that help minimize various challenges and thereby open up additional challenges and opportunities. What is typically preserved in this process are the traces of strategies that have allowed animals to navigate the specific challenges they have faced. So, we should proceed cautiously while also acknowledging that our cognitive capacities reflect the ways we are embedded in societies and have cultivated opportunities to make use of opioids in a wider range of contexts—put somewhat differently, we should acknowledge that it is the interaction between evolved tendencies and the structure of our socially shaped niches that is the source of our distinctive vulnerability to addiction.[2]

EPISTEMIC FORAGING AS SOCIAL COGNITION

Let's begin by considering what it means to highlight evolved sources of vulnerability. The first place you might think to look in this context is *On the Origin of Species*, Charles Darwin's masterwork, which attempts to explain how individual advantages gradually give rise to species with distinctive characteristics.[3] But we would like to start elsewhere. Darwin was also a keen observer of nature, and he drew upon the more classical Aristotelian traditions to catalog the ways that organisms grow and sustain themselves. He also explored the strategies that were being used by development-oriented botanists as they attempted to produce stronger agricultural strains.[4] Across a substantial body of work on orchids, carnivorous plants, fertilization, the motion of plants, and plant domestication, Darwin bolstered his claims about evolutionary processes, and he demonstrated that the practice of epistemic foraging could provide strong support for the hypothesis that the same processes are implicated in the evolution of plants and animals.

Since Darwin's time, data have continually accumulated in support of the hypothesis that plants are not nearly as different from animals as is commonly assumed. The earliest plants date to the origin of multicellular species, and plants have been evolving for at least a billion years. Like animals, they recognize and manage diverse challenges and opportunities, and they use many of the same chemical signaling systems as animals—including steroids, neurotransmitters, and peptides such as endorphins. According to current estimates, poppy plants have been expressing versions of opioid molecules for approximately eight million years, and there is suggestive evidence that the psychoactive properties of the opium poppy evolved in parallel with the emergence of opioid receptors in animals. These molecules have become pervasive features of diverse brains and bodies, and this has allowed pain regulation to become entangled with opium poppies across numerous species.[5]

The search for medicinal plants depends in part on widely shared cognitive capacities, and self-medication with bioactive plants has been observed across numerous species.[6] For example, gorillas and chimpanzees seek seasonally available plants to ameliorate the effects of parasites.[7] Orangutans have been known to seek pain-reducing plants to mitigate discomfort.[8] It is likely that the first use of poppies within our species is ancient. Many of our hominid ancestors were omnivorous, so they probably would have stumbled upon and ingested poppies to manage their experiences of pain. But to understand our unique vulnerability to opioid abuse, we need to look toward distinctive facts about human evolution and development.

By the time they reach maturity, typical humans have come to possess a complex suite of cognitive, affective, and attentional capacities that allow them to monitor events together, become attached to one another, and engage in acts of social- and self-regulation. But our physical capacities are incredibly limited at birth, and many of these limitations persist until relatively late in development. The slow rate of human development plays a crucial role in solidifying the social capacities that are distinctive of typical members of our species; specifically, it allows infants to spend a great deal more time learning from others.[9] Since human children are dramatically more dependent on caretakers than other big-brained animals, they can often observe a wider range of models as their cognitive capacities develop. As they do so, they learn that there are many ways that people engage with the challenges and opportunities they face. As a result, human infants learn what to pay attention to and remember from their caretakers, and this continually shapes the ways that they make sense of themselves and their relationships with others within their social world.[10] There is a great deal to say about our distinctively social capacities. But for our purposes, there are three things worth noting about the distinctively social features of human lifeways.

First, over our evolutionary history, humans have needed to survive and flourish in a range of environments very different from the savannas where the genus *Homo* first emerged. As they moved through these environments, human ancestors needed to explore the nutritional and medicinal value of various plants. Our closest nonhuman relatives, *Homo neanderthalensis*, also moved around quite a bit, but they appear to have found ways of living that worked, and they held on to them for numerous generations. They seem to have been far less concerned with technological innovation.[11] By contrast, early humans engaged in numerous forms of exploration and innovation as they pursued viability in an enormously wide range of ecological contexts.[12] We see traces of this history in the fact that many humans tend to be intrigued by new, alluring, and stimulating things. Of course, there are deep individual differences in willingness to pursue novelty. Moreover, the willingness to explore new options as opposed to exploiting existing strategies changes over the course of development—higher rates of exploration are typically observed earlier in life, and higher rates of exploiting existing strategies are more common later in life.[13] Finally, diverging developmental pathways, as well as variations in learning strategies, support a range of individual differences in what is explored and shape the likelihood that interest and creativity will expand our collective epistemic horizons.

Second, when humans learn about the world through observation, they do so in ways that bear the marks of our evolutionary history. Where environments are relatively stable, animals tend to be prepared to learn about the structure of their world.[14] This is part of the reason why dolphins rely more heavily on audition, while rats and mice rely more heavily on olfactory cues. Human engagement with the world also bears the marks of a complex evolutionary history. Typical humans *look* at the world, *watch* what others do, and *search together* for evidence of challenges and opportunities. To a large extent, the ways that we explore the world reflect the fact that most human populations have been highly mobile, migrating frequently to new contexts where old knowledge was unreliable. More than any other facet of our biology, the capacity to adapt to diverse contexts has been critical to our evolutionary success. This is why typical humans also display pronounced patterns of cognitive flexibility.[15] But flexibility always comes at a cost. We cannot be perfect optimizers; at best, we can be "good-enough" problem solvers.[16] We are vulnerable to misallocating resources under many conditions, and we tend to develop local strategies for dealing with difficult circumstances. Fortunately, this is not the end of our story!

Throughout our evolutionary history, humans have also exploited opportunities for cultural learning in ways that have resulted in capacities for managing a range of challenges and opportunities. It is worth dwelling on this point. The fact that typical humans possess a vast range of distinctive social capacities, including capacities for using language and technology, has allowed our species to explore the world together and share knowledge with people who have other forms of expertise. Rudimentary forms of communication and tool use were probably common among our evolutionary ancestors, and they persist in our primate cousins. However, the differences between humans and other primates are vast, and this reflects core features of the socioecological niches that our species has created and occupied. Put much too simply, typical humans can share knowledge with people who experience the world differently, allowing their explorations to become more fruitful and effective.[17] Throughout life, learning often depends on the knowledge that others possess. Typical humans communicate in ways that organize our social milieu, drawing upon shared epistemic resources to orient toward specific objects, learn about their properties, and group them in terms of their intended uses. This is possible because distinctive patterns of development allow typical members of our species to rapidly make sense of the behavior, intentions, and affective states of those with whom they share a social world. This is important to our understanding of opioid addiction because our social capacities have provided the basis for cultural understandings of what to eat, which medicines to use, and which substances to avoid.[18] We return to this point in the next section, and we return to this issue in more detail in chapter 4.

Third, and finally, the lack of social contact, especially positive social contact, can make people feel vulnerable and evoke feelings of discomfort. By contrast, interactions with others can placate the visceral discomfort of loneliness. Likewise, social laughter is often linked to a sense of well-being and often correlated with shifts in pain thresholds.[19] The story here is incredibly complex, but for now, the key thing to note is that various neuropeptides—such as oxytocin and vasopressin—play significant roles in facilitating social contact and withdrawal and in supporting feelings of social recognition and social support.[20] Furthermore, social touch can trigger the release of endogenous opioids, shaping feelings of social acceptance, and social laughter can trigger the release of endogenous opioids, placating pain and enhancing pleasurable feelings.[21] There is variability in each of these effects, and importantly, this is true of the roles that various neuropeptides play, as chemical signaling systems are calibrated against social regularities over the course of development.[22]

THE IMPORTANCE OF CATEGORIZATION

Over the course of this book, we will spend a great deal of time discussing these neural and chemical systems. For now, however, we want to focus more directly on the fact that many humans assign categorical labels to organize their thoughts and actions. We approach new situations by attempting to put them into the context of the categories we have developed from our past experiences. We employ taxonomies to discover new tools for thinking and acting together. Humans have often recognized "plants" as a subset of living things, and they have grouped them into clusters on the basis of their uses. The ease of categorizing plants has hidden mistakes, and we have needed to refine our taxonomies through the self-corrective inquiry of science. In the process, however, we have always foraged for coherence and some degree of control over the chaotic world we live in. This matters because how we categorize the opium poppy shapes its use as well as the likelihood that it will be abused.

There are many different ways for people to categorize a poppy. One person might treat it as a flower. Another might treat it as a food source. Another might group it with visually different plants that have noticeable effects on experience (such as coffee, chocolate, tobacco, peyote, and psychedelic mushrooms). Each of these ways of categorizing the poppy is socially shaped; each of them allows for generative inferences about what should be done in social contexts; and in every case, our understanding of the poppies as features of the world is shaped by interactions between our capacities to organize and categorize and the demands of navigating a social milieu where we must make sense of predictable and unpredictable events.[23] This is why typical humans generate ideas, test them, and revise them in light of social feedback. But we also build on interesting ideas and extend them in ways that support new strategies for organizing thought and action—and we do not just do this on our own.[24]

Diverse capacities for social engagement and shared attention, as well as the ability to share knowledge using language, also shape our understanding of the world. We articulate claims about complex networks of social relations, we coordinate complex social tasks, and we share elaborate forms of specialized knowledge. In each of these contexts, our interdependence is key, and it is likely that one of the primary reasons why observable differences in novelty-seeking persist within our species is that we can share knowledge, including knowledge about which plant can be eaten, which should be avoided, and which should be used for medicinal effects.

Here, too, there are evolutionary considerations in the background.[25] Human cognition has been shaped by a history of living in small cooperative groups, which required cooperation, social monitoring, and social support.[26] For most of our history, survival has been much less likely for humans who were neglected, and social contact has always played a crucial role in our ability to survive and flourish.[27] It should come as no surprise that substantial neural and cognitive resources are dedicated to managing the social milieu, including keeping track of who has done what.[28] Since using new plants in new climates meant that people could not rely upon shared wisdom as their only source of botanical knowledge, humans have needed to be active learners who can not only explore endlessly but also display high degrees of inertia and a willingness to search cautiously for new opportunities.

Understanding addiction requires an account of how these generally adaptive tendencies can be turned against individuals in specific ecological contexts, and this requires exploring the way that judgments, as well as the allocation of resources, are shaped by biases and misapplications of heuristic reasoning strategies. It also requires recognizing that our deeply social capacities affect opioid addiction in numerous ways, ranging from effects on the physiological response to pain to the search for patterns of social acceptance and the avoidance of social exclusion.

All of these things have mattered to our species. But it is important to remember that there is no linear story to be told about the evolution of our capacities. Patterns of evolutionary change are fragmentary and characterized by patterns of abrupt and continual change.[29] In the context of human action, however, patterns of stability and change often reflect the ways that we actively work to make the world conform to our needs and interests. This is not to deny that change has been a common feature of every environment we have inhabited. It has. But we have survived many of these changes because we are such a densely social species.[30] We have organized our societies and cultivated forms of teaching and learning that have minimized the impact of many ecological and social changes. Through collective activities and our shared memories—which are preserved in narratives and stories—we have been able to transform inhospitable terrains into habitable environments. Of course, new threats continue to arise. Old threats sometimes dissipate. And through it all, the poppy has softened the blow—and it will probably continue to do so.

Any account of how people have used the poppy will have to look to the wide range of different roles that this plant has played across different cultural traditions. The opium poppy and its chemical derivatives have come to saturate our world. But this plant has been with us for a long time, and its use has sometimes been more beneficial and sometimes been more detrimental to our well-being. We will explore some of these issues in chapters 4 and 5. For now, however, it is worth noting a couple of different ways that the categorization of opioids has shaped how they have been used and sometimes abused. The opium poppy is a natural object that can be eaten, used as medicine, and used to induce peaceful oblivion. It is also a beautiful flower. As we noted earlier, each of these factors has played a role in the ways that humans have organized and categorized this plant, made sense of its possibilities, and found ways of using it for good and for ill. But one thing to remember is that the expansive use of the poppy has required a high degree of control over the world.[31]

Humans have imposed various forms of stability on the world to make it available across times and cultural situations. One way this has been done is through gardening practices.

This mundane activity is a basic part of our exploratory nature. Charles Darwin was an avid proponent of testing and categorizing plants by their capabilities, often in the context of gardens. He was patient, thorough, and thoughtful, and he had a keen eye. Darwin was drawn, in his consideration of plants, to observations of the effects of light on the movement of plants as well as their coloring. The incipient science of botany that is reflected in Darwin's work aided our epistemic search. Foraging had naturally led to a desire to understand nature, but the new botanical sciences allowed knowledge about the natural world to flourish as never before. But there are other forms of epistemic foraging that are also found in gardens. Gardens represent the universal principles of nature in a particular way, and they open paths toward systematizing and cataloging.

Beyond these kinds of practices, the commerce in spices and herbs has long facilitated the spread of opium plants. At various points in human history, patterns of migration and colonization have helped the poppy to cross the oceans. This is the same process that was observed with a variety of medicinal and recreational plants (including coca, peyote, mescaline, and cannabis). In these contexts, the human tendency to "discover," cultivate, and share substances with others has sometimes taken a dark turn. For example, useful scientific discoveries have been commodified in ways that increase incentives to cultivate a plant beyond personal use. In some cases, the pursuit of scientific knowledge has even been shaped by the marketability of medicinal opium. This is possible because we are a social species and because of the complex ways we are bound to one another. The economic incentives that dominate the production and sale of opioids can often yield larger social incentives that push us toward more pervasive patterns of opioid addiction. This is something we should always remember as we think about the kinds of physiological processes that shape the experience of opioid addiction.

Another way members of our species have systematized the world is through art. This includes not only the aesthetic orientation toward gardens but also the ways we remember things when we hear them in music or read them in a poetic meter. This has long been a facet of human engagements with the poppy. For example, Homer mused on the allure of the plants of the gods, and he noted that they could mollify physical and psychological pain. But poetic visions run throughout human history, focusing on the cultivation of awe toward nature and its gifts. Across cultures, writers have offered literary lessons on addiction and romantic experiences of oblivion. We see similar depictions in painting, music (including the music of the 1960s), and even a Russian ballet, *La Bayadère*. This should come as no surprise. The opium poppy is one of the great wonders of nature. Humans probably found it while they were foraging for nutrients and healing strategies. But what they found was a miracle— a miracle with a dark side, which gave rise to abuse and overuse, in part driven by the pursuit of luxury and capability.

Finally, there are far more distant causal relations that organize and shape the broader character of our world and thus shape the likelihood of pursuing the experience of engaging with opioids. We can get a sense of this by thinking about the ways that our social world has been shaped to promote an endless drift toward—and beyond—the pursuit of comfort and pleasure (not for everyone, of course, and not in ways that are distributed evenly across race, class, and gender). For many people, the pursuit of comfort and pleasure has become a dominant motivation. But this motivation is also bent toward the distinctive challenges and opportunities that dominate our social world. So, some people continue eating popcorn when doing so is no longer pleasurable, and others give in to the pursuit of sweets where they are readily available. These tendencies are guided by the pursuit of pleasure but in ways that are shaped by advertising and the distribution of resources that play off of biological vulnerabilities. Likewise, some people become entangled with gambling when they are continually exposed to advertising that reminds them of possibilities that gambling—seems to—afford, especially where lottery tickets are readily available. Each of these phenomena might look superficially like forms of addiction because addiction is also a socially engineered exploitation of a biological vulnerability.[32] But being vulnerable to overeating in some contexts is not addiction, even if the exaggerated pursuit of sugar in nonnutritive contexts has aversive implications.[33] Likewise, the desire to gamble if the opportunity arises differs in substantial ways from the pursuit of opportunities to gamble where they aren't easily available. Put much too simply, addiction is never just a matter of overindulging when the opportunity arises. It is a form of cephalic dysregulation that bends patterns of thought and behavior toward the pursuit and use of opioids, and it does so across diverse contexts, including those contexts where opioids are not readily available. This is, at least partly, a result of how these substances shape neural responses to rewards and incentives.[34] But just as importantly, there are always situational and cultural factors that shape the pursuit of comfort and the avoidance of pain.

ANTICIPATION AND CEPHALIC DYSREGULATION

From a biological perspective, we should begin with the recognition that the brain is partly organized by anticipatory systems that sustain the pursuit of viability and support adaptive patterns of responding to diverse challenges and opportunities. These systems regulate the search for food, water, salt, and warmth, and they are also implicated in anticipating what will be useful to manage future contingencies.[35] Throughout our evolutionary history, humans have had to anticipate encounters with predators and the likely times and locations when food would be available. They have also needed to adjust their forms of risk aversion and risk tolerance in light of changes in the local environment, and they have needed to integrate new information into an overall understanding of the world. In each of these contexts, past events tend to shape present experience and the way that present experience is used to anticipate a range of possible outcomes. These forms of anticipation are often modeled using a statistical formula known as Bayes's theorem.[36]

The story is complex. Anticipatory strategies depend on the assumption that the future will look a lot like the past—and often, it does. In many cases, we respond to challenges and opportunities in ways that have worked previously, and we exploit neural processes that allow us to remember the distribution and value of previous rewards. But it is often impossible to predict the future since the future looks fundamentally different from the past. So, the brain must also continuously sample the information in its environment, evaluate predictions, and update them to preserve viability. In this process, the brain carries out active inferences that support forms of thought and behavior essential for learning about our world and pursuing viability.[37] Capturing these kinds of phenomena requires centering the fact that metabolic expenditure tends to increase when challenges and opportunities must be managed, and it requires highlighting the kinds of strategies the brain employs to anticipate likely challenges and opportunities while organizing behavior to minimize disruptions and metabolic expenditures.[38]

Of course, the world is messy, and changes in physiological and social needs are complex. So, these forms of anticipation are neither simple nor unified. Instead, various forms of anticipation are employed to support the pursuit of *viability* by producing patterns of thought and behavior that are sensitive to the likelihood of threats, rewards, and opportunities. In each case, we economize by assuming that the world and our needs will change in predictable ways. We rely upon diverse chemical signals, including endogenous opioids, to calibrate thought and behavior against the most salient features of our ecological and social niche.[39] But when access to a resource is disrupted or when our needs outrun our anticipations, we must find ways to learn more about the world, or we must start looking elsewhere for the resources that we need.[40] This is precisely where addiction has the potential to take hold.

At its core, addiction can be understood as a form of cephalic dysregulation, which is anchored to the brain and a cultural context. For example, the predicted reward of drug use often becomes tied to specific environments, making pursuit almost reflexive in these social and material spaces.[41] Positive memories become associated with opioid use, and the recall of such memories is then triggered in the schoolyard, backroom, alley, bar, coffee shop, or other salient locations. Sometimes, the allure of opioids can become so strong in these environments that thoughts and behaviors that are associated with the drug can be triggered by remote as well as recent memories. Often, such memories can begin to structure action, making prior events that led to a fix ultrasalient, thereby preparing a person to chase their drug of choice. Additionally, the memory of pleasure can become tied to specific activities *associated with* getting high, including chasing opioids, preparing to ingest them, and, in extreme cases, even experiencing withdrawal symptoms. In each of these cases, memories have the potential to crowd out other strategies for inducing positive feelings and fuel obsessive searching, preventing people from realizing the fleeting nature of hedonic responses.

The details will become clearer over the course of this book. For now, however, the main point is that the brain is a complex and multidimensional system that relies on *habits* to organize action and preserve our sensitivity to the salient features of our local ecology and social context. Diverse neurotransmitters and neuropeptides underlie the formation of habits and the organization of action, and diverse processes support different types of habit formation. Processes centered on a region known as the basal ganglia seem to support habit formation in cognitive and behavioral contexts.[42] Here, pulses of dopamine appear to modulate incentive salience and experiences of "wanting."[43] Incentive salience is often linked to experiences of pleasure and "liking," which are modulated by endogenous opioid systems. Importantly, habitual action is also shaped by activity in the amygdala, which appears to regulate appetitive search, and by processes that center on the nucleus accumbens, which seem to be associated with experiences of reward and anticipated reward in some cases. The collective activity of these processes, along with many others, shapes the cognitive and affective processes involved in habitual activity. This ranges from the moment of pleasure that is associated with listening to a familiar song or walking a beloved dog to more complex habits that are calibrated against the structure of our social and material environment. But critically, these systems also appear to collectively support the experience of addictive allure.

To get a sense of what this story might look like, consider the way that modern casinos are designed to exploit human vulnerabilities. The lighting, carpet, organization of space within a casino, and computational processes that are employed in modern gambling machines have all been designed to produce deeply immersive experiences.[44] These experiences aren't necessarily pleasant—people often report "losing themselves" in the activity of gambling. The effects of this are striking. Some people become so immersed that they learn to wear diapers when they go to casinos so that they can continue to play while they are urinating. Casino staff learn that they will need to clean a number of machines at the end of the night. In this context, patterns of thought and behavior are bent toward gambling, and the negative effects of doing so become far less salient than the pursuit of the immersive effects of this practice. The harmful effects of various addictions will differ across contexts, including the addictions to nicotine, gambling, and opioids. But there are common themes across these contexts: in each case, an object or a substance facilitates the capture of attention and shapes behavior in ways that are detrimental to human flourishing; and in each case, there are networks of social phenomena that shape and reinforce biological cravings in ways that give them a more determinate character.[45]

Opioids exacerbate these effects because they target the neural systems that are implicated in the hedonic aspects of craving. They also shape the experience of psychological pain, and they modulate relationships with other people. This includes the people with whom drugs have been taken as well as the activities that have been undertaken in the pursuit of opioids. When someone becomes addicted to opioids, compulsive pursuit, use, and dependence all become features of their biology. Their brain is altered, perhaps for a lifetime. Addiction *narrows* the range of phenomena that a person focuses on, yielding an obsessive anticipatory response to experiences that often recede with use. In part, this reflects large-scale shifts in the activity of processes that support learning, habit formation, and support the pursuit of reward and pleasure. In part, it reflects shifts in the impulse to learn and search, which become more narrowly focused, sometimes to the extent that addictive cravings are prioritized over needs for food, water, and shelter. The search for opioids can thus reorganize diverse anticipations, and as the experience of relief fades, so does the reward of *satisfying* the addictive craving. Thus, the craving for more *almost* seems to replace the pursuit of viability as the core biological demand.

As we argue throughout this book, chasing opioids is a perversion of the natural tendency to pursue viability. Searching for and consuming opioids can occupy and exhaust the capacities that would typically sustain the pursuit of viability, leaving few resources to navigate other challenges or seek out other opportunities. It is important to remember, however, that opioid addiction differs from most other forms of habitual thought and behavior. Opioids crowd out other habits, allowing predicted rewards to become more potent and experiences of satisfaction to become much too brief.[46] When this happens, people often try to prolong the reward by seeking bigger doses or stronger drugs. This can lead to an increased tolerance, driven by a seductive and dangerous attempt to pursue an unachievable sensation.[47] But it can also yield isolating experiences, given the impact on embodiment, action, social engagement, and engagement in the larger culture. This point is key: the *biological* changes evoked by opioids always interact with the structure of the *social* and *ecological* world. It is true that opioids change the brain. But as addiction is molded into the brain, every attempt to make sense of the world and integrate new information is changed. This is why it would be a mistake to claim that addiction is a "brain disease."[48] Addiction does not simply disrupt neural activity in the way that infections disrupt lung activity. Instead, the brain is continually shaped by a social and ecological context that is shared with others.

WHY DO PEOPLE USE OPIOIDS?

Throughout history, people have used opioids for many different reasons, but the pursuit of pleasure is rarely primary. Of course, some people are motivated by thrill-seeking and the pursuit of pleasure, but many hope to loosen the grip of reality or retreat from social interactions. Many seek smoother engagements with music or smoother forms of social movement. Many want to manage pain. Across all of these contexts, opioids can decrease physical or psychological discomfort and minimize various kinds of anxieties. But the use of opioids is always a drastic remedy. Opioids dull pains caused by injury, arthritis, and surgical wounds—but this can often lead to ever-increasing doses of drugs like oxycodone. Moreover, opioids can dull the pains of social anxiety and placate the pain of physical or economic insecurity. They can yield experiences of "isolated perfection," where pains, worries, and psychological exhaustion dissipate and additional causes of stress are minimized or eliminated. This is not, however, a stable state, and the allure of peaceful oblivion often gives way to a craving for opioids and a need to alleviate the discomfort of addiction. Some people spend a lifetime chasing the feeling of inner peace from their initial use of opioids, but they rarely succeed in recapturing this feeling. Instead, they confront pervasive discomfort and dysphoria and newfound desires to relieve these forms of distress.

In some contexts, the slip from use to abuse often turns on a myth about past experiences, attempts to recover those past experiences, and obscuration of the painful reality of increasing discomfort. Addiction to opioids can often yield endless cycles of craving, searching, waiting, hurting, and seeking oblivion. However, the initial desire to try a substance like heroin might reflect a search for thrills, novelty, or peaceful oblivion, and the pursuit of this substance might become a full-time occupation as the desire to use it becomes stronger and stronger. Increasing use often yields a greater need; greater need often supports increasing appetites; and as appetites grow, other desires wilt away, allowing the pursuit of opioids to become a primary motivation supported by a pervasive appetite. The result is a life dominated by appetitive search, with the satisfaction of a waking dream becoming nothing more than a distant memory. But this isn't the end of the story. New forms of dopesickness must be navigated, which include additional kinds of cravings, angst, sickness, backaches, shivers, pain, insomnia, vomiting, and more. This form of pain often emerges when someone tries to stop using opioids, and it often leads to the re-entrenchment of use, driven by the sensible desire to avoid the uncomfortable and world-destroying experience of withdrawal.

Similar processes are observed with dependence on pain-reducing opioids. This is part of what makes the current opioid crisis so frightening. There is no free ride when it comes to managing pain, and the initial hope of managing pain can be the beginning of a long travail. To understand why, we must look at the interactions between physiological, psychological, and social factors that produce and sustain opioid addiction. The story is complex. We focus primarily on opioids in this book, but opioid addiction is often accompanied by the abuse of other substances. Some people assume that addiction is a biological drive or a pernicious personality trait. Such perspectives downplay the complexity of addiction. Unlike an infant who is born with physical dependence, the process of *becoming* addicted entails a shift in focus and an increasing vulnerability to abusing opioids. This process unfolds through the internalization of new motivations and the development of a global shift in the structure of experience. Such changes are rooted in biology and capacities for habit formation. As we argued above, human cognition is rooted in biological and cultural evolution. At some point deep in our evolutionary history, we found opium, discovered that it could relieve pain, and began to use it. Like other species, we return to plants that placated discomfort. But we learned to cultivate poppies. We developed knowledge practices (and commercial practices) that centered on this plant and its derivatives, and our cognitive and affective capabilities made us vulnerable to the misuse of these addictive substances because they allowed us to push beyond their adaptive use and discover ways of using opioids that have reduced the capacity for living a good life.

The complexities of this situation are the primary reason why a plausible perspective on opioid use and abuse must be situated against the backdrop of differential access to social assistance and the impediments that unfold in the context of racial, ethnic, economic, and gender inequities.[49] This is no easy task. After all, the vulnerability to addiction is shaped by a complex and heterogeneous array of cultural factors and a playing field that has never been fair or equitable.[50] Our aim in this book is to clarify some of the interactions between biology and culture that shape opioid use and abuse and to clarify what it means to claim that the vulnerability to opioid addiction is an effect of our biological and cultural evolution. But it's hard to know exactly where to start.

DEVELOPMENTAL COMPLEXITY

In typical and pathological contexts, people try to strike a balance between desired behaviors and available resources. Over the course of development, people often learn to prioritize certain actions and manage diverse needs and interests in ways that facilitate survival and—where possible—flourishing. Addiction can trap a person in a situation where bad judgments prevail, and behavioral inhibition is less likely to support the avoidance of lethal and harmful threats. But to understand why this is the case, we must consider the complex networks of interaction between physiological and social contexts, including those that change over the course of development. Adolescence is commonly noted as a potential breeding ground for opioid addiction. Increasing autonomy and social demands shape many of the ways that adolescents navigate and learn about the world.[51] People often display heightened levels of risk tolerance in adolescence, leading to more open-ended explorations of diverse experiences; and various kinds of drug use often occur during this period, ranging from explorations to intermittent and sustained uses.

Strikingly, fewer than 20 percent of nonprescription opioid users will end up addicted to opioids.[52] Where addiction does take hold, it can reduce inhibitory control, reflecting a shift away from cortical processes and toward subcortical networks centered on the amygdala and bed nucleus of the stria terminalis. These changes in neural activity appear to increase the vulnerability to addiction—and they are sometimes observed during adolescence.[53] But even here, there are porous relationships between thought and action and the relationship between what happens in the head and what happens in the world. This makes it difficult to disentangle the biological and social causes of addiction. To be clear, however, this is not a reflection of our current ignorance; it is the result of the fluid and deeply entangled nature of mind, body, and world. There are two things to notice here.

First, opioid abuse is commonly accompanied by the abuse of other substances, such as alcohol, tobacco, cocaine, and pills. But this is unlikely to reflect a personality trait or a biological disposition that leads people toward addiction. In many cases, such pursuits seem to reflect corrective measures or attempts to regain traction on experience as addiction begins to take hold.[54] Second, it is hard to understand how someone who is terrified of needles can find themselves shooting up with heroin; it is hard to imagine how someone can accept living in dangerous situations or squalid and abandoned buildings; and it is hard to understand the willingness to face the harms that arise in the wake of addiction, including the threat of incarceration and subsequent violence and the necessity of abandoning human connections to survive. But these are common effects in the context of opioid addiction.

The power of opium to produce aversive effects has been well-known for a very long time, and, like many behaviors that seem unfathomable under typical conditions, opioid abuse often arises for intelligible reasons—a desire to escape something awful, ameliorate discomfort, feel less, sink into oblivion, feel fearless, or fit in. Some people turn to opioids to forget possibilities that have dissipated. Some people seek a sense of euphoria.[55] Some seek an escape "from a sense of insupportable loneliness and a dread of some strange impending doom."[56] Some people are simply motivated to avoid pain but fall prey to medical overprescription.[57] Someone who suffers a football injury might find that opioids also minimize their experience of psychological and social discomfort. As they seek the amelioration of psychological discomfort, addiction might take hold in ways that are easy to understand but difficult to predict.

Often, novel forms of sickness and suffering will emerge alongside the experiences of loss and discomfort, yielding experiences that are commonly known as withdrawal. This might lead to the destruction of families as people become embroiled in the use of opioids. Sometimes, people's families will continue to care about them and care for them, even as the experience of isolation increases. Opioid use can also become a family lifestyle, with children and adults both becoming addicted. Older individuals who are introduced to opioids as prescription medications start to face addiction. While they might not seek heroin, their experience is bent toward the pursuit of opioids. Likewise, young athletes who are treated for injuries are increasingly likely to face addiction, and their parents are forced to watch the downward trajectory of their vulnerable children.

Across times and cultures, engagements with opioids have taken numerous forms, and people have coped with the challenge of opioid addiction in diverse ways that reflect their social and cultural locations. But the first experience of opioids can yield a seemingly harmless and warm high, with little indication that the allure of opioids can become an overwhelming motivation. This is a theme that is ripe for exploitation by businesses that can capitalize on cravings.[58] And as it has become more common for thought and behavior to be bent toward opioids, numerous markets—legal or illegal—have arisen to meet the high demand.[59] These phenomena are part of a narrative that is becoming more deeply familiar. It is important to remember that as more middle-class and affluent kids are afflicted, an increase in concern about the opioid crisis has followed. Unfortunately, this pattern of concern tracks the emergence of cultural and regional differences in the method of using opioids, and this is something to keep in mind as you read the rest of this book.

NO ONE ESCAPES

In this introduction, we have suggested that our evolutionary history is tied to two facts about opioids: they can be used to treat pain, and we are vulnerable to abusing them because of how they affect us. That said, some people never become addicted, and others successfully pursue sobriety after becoming addicted. Perhaps this reflects differences in the experience of pleasure and pain. We all have such experiences, but they differ in innumerable ways. With opioids, the dissipation of pain can often be rapid. For some people, this experience can be enjoyable. Addiction might be more likely where the absence of physical and psychological discomfort becomes a defining feature of pleasant experiences. But many people initially dislike the experience of the opioid high. Indeed, many people feel nauseous the first time they use opioids. Likewise, some people report being afraid of needles and then drifting toward needle use as they become addicted. Finally, the experience of the opioid high becomes an acquired taste for some people.[60] But not everyone who experiments with opioids becomes addicted. There must be some biological factors at play to support these differences. At present, no one really knows why one person becomes addicted to opioids while another person does not.

There might not be any single type of addictive personality nor any single social location that contributes to addiction. So, any story we tell about these diverse factors must be careful to avoid making illicit assumptions about what makes someone the wrong kind of addict—this is one of our core claims in this book. Throughout this introduction, we have offered some of the background that will shape our claims in the remainder of this book, and we will thus begin to explore some of the ways that physiological and psychological constraints interact to shape the pursuit and use of opioids. The key thing to remember is that there are numerous complexities involved in the experience of opioid addiction. The drive to alleviate withdrawal, obsession, and narrowing of focus can become pervasive in experience.[61]

The feeling of comfort can depart and be replaced by an obsession. Over time, higher doses of the drug become necessary, and the pain of withdrawal tends to increase. The result is an endless obsession with finding and using opioids by any means necessary. However, this is rarely where the story ends. These effects always reflect complex relationships between brains, neurotransmitters, and social contexts. Put somewhat differently, addiction is always a physiological, psychosocial, and existential phenomenon, and the effects of addiction can unfold across numerous different dimensions, including factors that are shaped by social status.

In most cases, however, tolerance will tend to increase with use, meaning that more of the drug is necessary to achieve the same effects.[62] Moreover, as tolerance increases, both perception and motivation will tend to be *bent* toward an internalized need to consume more. This often yields an increase in the rate of psychobiological aging. Even under the best circumstances, human tissues and organs decay, but overuse can yield a more rapid breakdown of functioning. In part, this is because cortisol is elevated during the pursuit and consumption of opioids,[63] and chronically elevated cortisol levels accelerate the aging of tissue, breakdown of immune function, and onset of old age.[64] But critically, these effects are also shaped by the distribution of care and support and differences in access to treatment facilities, which are never equitable. This is part of the reason why differences in the physiological and cognitive effects of addiction are never evenly distributed. Some people are more likely to be incarcerated. Some are more likely to pursue opioids in ways that require using dirty needles. A lot of people are more likely to die from overdoses because the heroin they have access to is cut with fentanyl. However, these are not reflections of differences in intelligence or capability. They reflect differences in which resources are available, patterns of rampant paternalism, and ethnic and racial disparities and unfair sensibilities and practices.[65] What we hope to promote in this book is an understanding of how these kinds of social features interact with our shared vulnerabilities. But first, we need to say quite a bit more about the nature of our vulnerabilities.

In chapter 1, we turn to questions about the nature of habits and the ways they shape engagements with opioids. As we explore this story, we focus on the importance of effort and decision making in typical and addictive contexts. From here, we dive more deeply into the biological processes that shape and sustain addictive cravings and the experience of addiction. In chapter 2, we turn to a more detailed discussion of our current understanding of the brain's endorphins and the roles that these and other peptides play in a broad range of strategies for managing various challenges and opportunities. In chapter 3, we examine the experience of angst and despair that commonly arise in the context of addiction, again focusing on our current understanding of the role of biological processes that shape such experiences. With this background in place, we then begin to explore some of the diverse ways that social factors have shaped our understanding of opioid use and addiction. In the last two chapters, we turn to questions about the management and treatment of pain by healthcare professionals and pharmaceutical companies, and we explore the nature of agency and recovery in light of the complex networks of biological, psychological, and social constraints that we have discussed throughout the book.

1

EFFORT AND DECISION MAKING

Widespread patterns of addiction pull at many of the threads of our social fabric at once. The most recent National Survey on Drug Use and Health estimated that over a million people in the United States used heroin in 2021, and that half of those also misused other opioids.[1] The actual numbers are probably higher. Some people turn to opioids to escape psychological or physical pain; others pursue the thrill of exploration. Some write glowingly about opioid use, exposing others to the danger of addiction while never becoming addicted themselves. For example, Allan Ginsburg famously wrote movingly about using heroin and then moved on. By contrast, some become trapped in the haze of addiction for a long time or suffer far worse fates. The comedian Lenny Bruce, for example, pushed the boundaries of social critique but died from a heroin overdose, soaked in blood and despair. Finally, there are people who use opioids responsibly and still slip toward addiction. The bioethicist Travis Rieder, for example, struggled through withdrawal after numerous doctors made complex pain management decisions that left him dependent on opioids.[2]

The patterns of addiction and subsequent breakdowns in capacities have become part of our shared cultural knowledge. But it remains less clear what drives destructive spirals, and it remains difficult to predict precisely who will suffer from addiction.[3] This situation is neither new nor unique.[4] Opium use has a long history, organized by medicinal use and ritual significance, and the vulnerability to abuse has come into view many times before. During the Victorian era, doctors attempted to mollify pain, but they enabled broader patterns of addiction.[5] Something similar happened in the United States during the nineteenth century, as our first opioid crisis unfolded in the wake of the Civil War.[6] In our current context, it is often difficult to counteract the allure of profits. Often, people try to make the right decisions, but find that opioids gain a toehold through medical prescriptions. Larger quantities of opioids are then distributed to an addicted populace because there is money to be made. While many people resisted, for example, the distribution of oxytocin was deeply entangled with the monetary interests of the Sackler family. These contexts are all unique. But in each of them, the entanglement of culture and biology has shaped how addiction takes hold.

While we focus on biological factors in this chapter, it is important to remember that addiction is not caused by biology— it is a form of experience that occurs in a social context in ways that are shaped by individual and social histories. Many people lack access to meaningful alternatives to opioids for pain relief. These same people often have reduced opportunities to control how their lives unfold and an increased need to seek an *escape* from pain and hopelessness. Some people use opioids to escape social trauma and eventually become addicted. But not everyone does—this is where biological factors matter. Some people are less able to tolerate pain, some people have predispositions toward addiction, and some people find it more difficult to break habits. In this chapter, we consider some of the factors involved in the appetitive search and some of the ways they can bend action and experience toward the pursuit of opioids.

EFFORT AND HABIT

It is often assumed that opioid addiction occurs when initial use gives way to occasional use, eventually leading to obsession and compulsion.[7] This assumption highlights a loss of control and the development of impulsive habits. On this basis, it is sometimes claimed that addiction reflects a change in neural activity such that cortical control is replaced by subcortical processing that operates more habitually.[8] This perspective assumes that goals are freely chosen while habits operate outside of cognitive control. However, goal pursuit, habit acquisition, and learning are cognitively and neurally intertwined. Both habits and goal-directed cognition are necessary to preserve the range of flexibility required for navigating a complex and multidimensional world.

Habits are structured actions that become part of a person's repertoire for coping with common challenges and opportunities. As such, they play a substantial role in adaptive thought and behavior.[9] So, it shouldn't be a surprise that they are not localized to a particular brain region. Instead, they depend on numerous neural processes that are distributed throughout the brain.[10] Together with goal-directed cognition, habits allow us to chart alternative courses of action. They facilitate rapid appraisals of objects and events and rapid inferences about how and when to act. Of course, they can become insensitive to change. But where they do, this often reflects their centrality to ongoing appraisals of salience and the necessity of minimizing neural and cognitive expenditure.

Goal pursuit often depends on habitual forms of action, which reflect what we have learned about specific contexts and what we can do when specific possibilities come into view.[11] Furthermore, goal-directed action can unfold almost automatically when it is supported by robust forms of incentive salience.[12] For example, opioid search is often shaped by habits, which are prefigured by entrenched forms of incentive salience and a narrow range of perceived possibilities for gaining access to drugs. But this yields patterns of goal setting and deliberation that are shifted toward seeking, finding, and using drugs. In this context, processes that would typically support adaptive behavior are bent toward pursuing, acquiring, and using opioids. But this is not because freely chosen goals are highjacked—it is because habitual and goal-directed behavior have been bent toward these possibilities. In the context of opioid addiction, this can sustain the compulsions that allow opioid pursuit to take precedence over other needs and interests.[13] But this is not just a shift in habits—it is a fundamental transformation of the relations of salience, effort, and goal-directed behavior that organize the ability to cope with diverse challenges and opportunities.

Many processes that organize thought and behavior depend on dopaminergic activation, and many are shaped by the habits that pervade our lives.[14] We return to this point below. For now, we mention this fact only to soften the assumption that habits are distinct from goal-directed thought. Habits are sometimes thought to occur in the absence of choice.[15] But the pursuit of goals shapes neural activity even when action is compulsive.[16] When a rat bites at an empty spout to gain access to opioids or when cravings persist in the absence of reinforcement, this reflects the fact that they cannot achieve their goal. The rat is frustrated and fixated, and their capacities for adaptive learning have been compromised by the entrenchment and stabilization of an unachievable goal. Our intentions, plans, and decisions are more deliberate. But here, too, choice is a matter of degree. Many choices are contextually dependent. And our habits, good or bad, orient us toward salient opportunities for action.

It might seem clear that freely willed actions are chosen deliberately. But, as William James notes, *willing* depends on effort, which is enacted throughout the body to support action, appearing "in every muscle as irritability."[17] From a neuroscientific perspective, it is less clear that willing occurs in ways that are independent of habit. Effort is often necessary when there is a conflict among choices.[18] It is often necessary to avoid harm and procure desirable things,[19] and it often reveals difficulties inherent in self-regulation.[20] Put bluntly, effort will be present whenever self-control is necessary, and both effort and choice will always depend upon the *motor* systems that guide behavior.[21] This doesn't mean you must move to think, but it is a deep fact about neural design that cognition is integrated with abilities for actively navigating challenges and opportunities.[22] This is why the neural processes that support exploration and movement are deeply entangled with those that sustain cognition, including habits, and capacities to deliberate.[23]

Finally, and perhaps most importantly, effort pervades the experience of addiction. Moreover, irritability becomes the plight of many people who are addicted to opioids. They strain against addiction. They strain to maintain access to drugs. They strain to get a foothold on the opportunities that would lead them out of addiction. These forms of effort might be more obvious in the context of addiction, but they reflect general features of cognition, which are bent toward the pursuit, acquisition, and ingestion of opioids. Opioids do not sidestep or hijack adaptive capacities to deliberate, decide, and will—they shape these capacities. Moreover, the habits associated with addiction are not significantly different from other kinds of habits. In each case, effort is required to resist habitual patterns of thought and action, and challenges will often arise in the context of self-regulation.[24] This is why it is a mistake to treat addiction as a disease of the will or a personal failing—addiction is a way that decision-making faculties can lead to actions and ways of exerting effort that contradict a person's best interests.

BENDING NEURAL PROCESSING

In the introduction, we suggested that addiction *bends* neural processing toward the pursuit and consumption of opioids. But what does this mean? Recall our claim that addiction is not just a biological hunger, though it relies on similar neural circuitry. Sodium craving reorganizes neuronal expression in the nucleus accumbens to reflect an expanding salience of sodium-relevant events.[25] Something similar occurs in the context of drug use.[26] In each case, anticipated needs support the experience of craving. Such anticipations are the forerunner of voluntary action.[27] They are the source of the "goals" that an active brain employs in pursuit of good or bad objects.[28] So, we might say that neural processing is bent toward pursuing opioids when opioids are the source of various cravings, anticipations, and goals and when opioids are pursued even where other challenges or opportunities are salient.

Experimental models in animals suggest that as addiction progresses, animals become less sensitive to the relationship between their actions and opioid acquisition. They also become less sensitive to information that would extinguish the association between action and opioid acquisition. For example, they hopelessly lick at an empty spout in hopes of ingesting opioids, and they do so until they are exhausted.[29] This suggests that the ability to abandon unsuccessful behavior is degraded and less sensitive to available information. But diminished flexibility does not mean goals are no longer at play. Still, it is important to notice that addictive behavior often reflects habitual forms of incentive salience, which are supported by subcortical processes that center on the basal ganglia. Here, a neurotransmitter known as dopamine (figure 1.1) seems to play a crucial role in shaping cognition and behavior.[30] But we must proceed cautiously, as the roles that dopamine plays are often misunderstood.

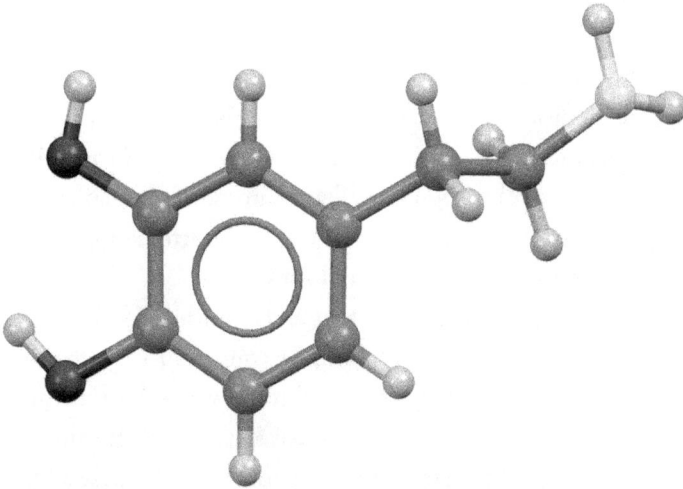

FIGURE 1.1 A dopamine molecule. Wikimedia Commons.
Public domain image.

Neurotransmitters like dopamine play numerous roles in the organization of cognition and action. It is sometimes claimed that dopamine is a pleasure molecule or a reward molecule. But it is crucial to the organization of action and motor control. In different contexts, it can function as a neurotransmitter or a diffuse chemical signal that boosts or suppresses synaptic signaling, depending on where it binds to a receptor. This matters because dopamine doesn't have a single function, but its interactions with a diffuse range of neural systems allow it to shape cognition and action in numerous different ways. Dopamine affects everything from motor behavior to impulsivity and motivation in ways that can be observed in the context of numerous psychological challenges.[31] Furthermore, dopamine depletion can affect task performance and incentive salience even if it doesn't affect learning. It can also enhance the salience of specific objects or events, even where they are not rewarding.[32] Finally, dopamine sometimes appears to play a significant role in learning the causal structure of events and learning the associations between different objects.[33] The key thing to notice here is that dopamine is a fundamental neurotransmitter that plays adaptive as well as nonadaptive roles in cognition. This neurotransmitter appears to play crucial roles in many of the activities surrounding opioid pursuit and the lifestyle of use and abuse.

Sometimes, it is suggested that addiction hijacks the dopaminergic reward system, diminishing control in ways that are analogous to a takeover by a hostile force. Perhaps this is because addiction narrows attentional focus, even though opioids become less rewarding over time, or perhaps it is because addiction shapes a person's orientation toward events and opportunities. With continued abuse, this can *feel* like the brain has been hijacked, as the awareness of available options is diminished and various forms of irrationality become more common. But these kinds of behavior occur in specific social and material contexts. They occur in contexts where attention is bent toward the pursuit of opioids and in contexts where the possibilities that the world affords are organized by changes to habits and patterns of perceived salience. All of this matters because the effects of dopamine regulation on cognition are pervasive.[34] Dopamine is released while ingesting opioids.[35] It is also tied to incentive salience in ways that shape appetitive behavior and the exertion of effort in pursuit of opioids.[36] Regions of the brain that are tied to this neurotransmitter are activated when people ingest opioids—and likely when they imagine ingesting them as well, as dopamine seems to be closely tied to both the salience of experiences and anticipation of experiences. Put much too simply, dopamine plays a crucial role in all forms of motivated behavior.

Importantly, reduced levels of dopamine are also observed in the context of drug use when the experience of control is reduced and impulsivity is increased.[37] This converges with evidence that dopamine regulation plays a crucial role in behavioral inhibition. Here, the most revealing evidence comes not from studies of addiction but from the metabolic disease known as phenylketonuria (PKU). PKU modulates dopamine availability in the brain, resulting in reduced levels of behavioral inhibition and patterns of delayed attention, which shape learning throughout life.[38] Of course, there are numerous complexities involved in a distinct developmental pathway like this. So, we shouldn't attempt to draw inferences about typical or addictive cognition from this case. However, there is suggestive evidence that genetic contributions to a person's temperament impact the likelihood of substance use and abuse.[39] These effects appear to at least partly derive from the impact on the dopamine-associated search for novelty.[40]

ACTING AGAINST
OUR BETTER JUDGMENT

The existing data suggests that dopamine plays numerous roles in regulating the awareness of opportunities for action and the perception of salient possibilities. This is a double-edged sword. Where enhanced salience sustains a search for beneficial resources, it can be adaptive. Where it amplifies the allure of substances like opioids, it can undermine abilities to resist options that are less desirable. But mastery and skill are the basic stuff of life. We try to master tools, musical instruments, sports, social and aesthetic sensibilities, and more, and dopamine is likely to play a crucial role as we resist worse options to pursue better ones, disengage from bad habits before they become entrenched, and track the salience of available and alternative possibilities in each of these contexts. Critical reflections on these facets of human psychology have been ongoing since at least the time of the Greek and Sanskrit epics. Aristotle famously uses the term *akrasia* to characterize the tendency to pursue a less desirable option when there is a more desirable option available. He viewed the lack of mastery, or the lack of control over how our actions unfold, as one of the key faults of human psychology that we must work hard to avoid—and there is something deeply right about this claim. But it is important to remember that people make flawed decisions even under the best conditions, in part because their reflective desires compete with diverse additional needs as well as drives that have become embodied in neural processing.[41]

We can navigate challenges that we expect to encounter with cognitive systems that have been calibrated through learning. But these systems are far from perfect, and our problem-solving and decision-making capacities are far from exemplary, even under the best of conditions. They are bound to the contexts where they have operated, and in different situations or different emotional contexts, we might come closer to—or fall further from—ideal forms of reasoning. This is probably inevitable since most contexts where we must make decisions are layered with ambiguity. But this means that everyone will sometimes act in ways that undermine their interests—either because they lack relevant information or because their attention is bent toward possibilities that lead them in the wrong direction. Of course, we can sometimes delay satisfaction to pursue longer-term interests. But the allure of short-term satisfaction and immediate reward can compromise this capacity, and we are bad at recognizing how bad we actually are at delaying satisfaction.[42] Our mistaken views of ourselves have been explored using diverse experiments in behavioral economics, but the key thing to notice is that we often misframe the context of choice, and we are often misled by forms of statistical inference.[43] Fortunately, we can often get by *well enough*, by relying on evolved and learned heuristics, but here, too, we are vulnerable to mistakes where our heuristics do not apply.[44]

These limitations are all exacerbated in the context of addiction. Opioids impact all of the capacities we have for problem solving and appetitive decision making. The processes that facilitate decision making can be bent toward the pursuit of opioids, and the allure of opioids can make it feel as if the need to pursue them is beyond the bounds of control.[45] But this happens because addiction magnifies some of the darker facets of human psychology, sidelining some tendencies for rational reflection and foregrounding worse options when better ones are readily available. This can lead someone to devote all of their deliberative capacities toward ends that are self-defeating. Because addiction changes how a person orients toward salient possibilities, it can dramatically reduce capacities for using heuristic strategies to make better choices. As a result, the pursuit of salient possibilities can be distorted by an increased motivation to pursue a salient and addictive substance.

Just as importantly, these effects are also exacerbated where local environments obscure the negative consequences of pursuing opioids or heighten the forms of psychological and physical pain that support a craving for escape. These effects are not life-sustaining, and they will often diverge from some of the larger goals that an individual attempts to pursue—but the salience of opioids can come to support an all-consuming motivation, which is pursued at the expense of everything else that might matter. It is irrational to ingest opioids for pleasure, and pursuing them in the context of addiction might seem profoundly irrational. But rationality is not the issue. Opioids bend thought and deliberation in ways that restructure goal-directedness as well as habitual action. In doing so, they change which options are available and which options should be pursued given the social and physiological constraints of addiction.

SOCIAL DETERMINANTS OF CHOICE

These claims about habit and decision making are far from the end of the story about how choices can become compromised in the context of addiction. Our evolutionary history provides us with capacities for pursuing a wider range of options than is observed in other animals. It also supports a wider range of strategies for accessing and using opioids. Likewise, processes of cultural evolution have provided numerous tools and skills, which dramatically expand the range of available choices. As a result, we inhabit a world that is organized by responsibility practices, diverse opportunities for self-determination, and numerous possibilities for social engagement. Unfortunately, in our unequal societies, many more paths are open to those with resources, and many opportunities are closed off for those who are socially and economically marginalized. This matters, as social groups can cause the kinds of pain that individuals seek to escape.

These factors rarely play a significant role in accounts of addiction that are developed and tested in laboratory environments. There are few available options in laboratory contexts. So, the goal-directed behavior observed in such contexts is constrained. Such research clarifies the neurobiological processes that shape and organize goal-directed behavior, but models of addiction based on this research often leave out the role of our

material and cultural environments in organizing addictive cravings and the pursuit of opioids. This matters because choices
that support locally adaptive behavior often become solidified
in action, repeated actions often become habits, and thoughts
and decisions are often anchored to facts about our material and
social environment. In the context of opioid addiction, interactions with specific people and places will often shape the way
that addiction unfolds. This, too, is a double-edged sword, as one
way to tame future behavior is by *limiting* current choices.[46] But
the attempt to get clean will often require struggling against the
groups we are entangled within, and this can often evoke further
experiences of conflict or existential angst, which keep a person
bound to the pursuit of opioids—we explore these issues in more
detail in the remaining chapters of this book.

All of this said, since habits narrow our choices and set us on
a certain path, the strategic development of habits can sometimes reduce the allure of addictive substances. But this is no
easy task. Consider the ways that precommitments can reduce
available options, impose costs on relapsing, and enhance the
reward of avoiding use and abuse. Like Ulysses's decision to bind
himself to the ship's mast to resist the siren's call, precommitments constrain us and make it easier to make the right choice
than to make the wrong one.[47] In the context of addiction, this
requires developing a commitment to avoid substances of abuse
and the people and places that support vulnerability to relapsing. This is likely to recruit neural systems that support the pursuit of specific habits and longer-term interests.[48] It is also likely
to recruit systems that support threat avoidance and defensive
behavior, as knowing *what to avoid* and practicing thoughtful
avoidance are often critical for recovery.[49] But even with these
strategies in place, the cultivation of new habits and the avoidance of potential triggers is difficult. Effort and rationality will

always be required to stay the course since we are rarely *literally* bound to a mast.

Sadly, opioid pursuit can often overwhelm even the most promising precommitments. In part, this is because it is difficult to avoid every potential relapse trigger, so thought and behavior are often bent back toward opioid pursuit. There are useful heuristics that can support the effectiveness of precommitments— for example, avoiding *this place, that person,* or *those situations.* But precommitments can be undermined by competing needs and desires, and, in the context of opioid addiction, this situation is complicated by the fact that imagination and memory can trigger many of the same processes as perceptual experience.[50] Put much too simply, anything that might trigger memories or plans associated with using or pursuing opioids can reawaken the addictive cycle—and in each case, dopamine will be recruited by the reawakening of the salience of opioids. Intriguingly, however, pharmacological blockades of the dopamine D3 receptor can reverse the escalated use of oxycodone.[51] But often, this option will not be available, and it may not even be effective in every context. But what other options are available?

To begin with, it is important to remember that self-binding behaviors need not be things that individuals do on their own. Consider the kinds of tasks that Narcotics Anonymous suggests for individuals: making coffee for the group, committing to walks together, committing to things that create links between specific events and actions.[52] By working together, people can sometimes find ways of binding themselves to shared behaviors, which scaffold the ability to limit exposure to situations where they are vulnerable. Such group-binding can provide the kinds of stability that support existential well-being, at least for some people. Though rationality is constrained in significant ways by addiction, there are often ways for people to work together to support and guide action, relying on interactions between social monitoring and self-monitoring to stabilize new kinds of habits.

This works best when people share strategies for holding one another accountable, which are grounded in ideals that everyone in the group has adopted.[53] This helps to preserve autonomy over shared actions, and it supports community-centered standards for praising and sanctioning one another. Over time, this can even provide a context where new habits can be constructed, as habits reflect our repeated activities, including our activities within the groups that we identify strongly with. But far more significantly, this makes it possible to build shared expectations, which allow people to receive evaluative feedback from others that will support better ways of living and acting. Finally, as new habits stabilize, shared expectations can begin to shift toward more rational patterns of activity, yielding a socially scaffolded form of freedom that can be entrenched through shared practices.[54]

There are better and worse ways for this to happen. People who struggle against addiction are no longer treated as "disease carriers" who should live outside of society.[55] Nor are they assumed to have a "morbid craving for morphia."[56] But often, the public response to opioid use and abuse is to make it illegal and to incarcerate those who use opioids. This method typically fails if the goal is to keep opioids away from those who pursue them. In fact, rehabilitation programs, prisons, asylums, institutions, and many other locations can become key features of a person's life, which further entrench their habits and incentives and provide new contexts for acquiring and using opioids. This isn't inevitable, but people learn to engage in different kinds of performances, and they develop patterns of attention and salience that are linked to different contexts.[57]

Adapting to social contexts is an essential aspect of our evolved inheritance, and some ways of adapting to institutions will lead to the maintenance of addiction, while others will help people overcome it. This matters because even the most progressive approaches to addiction, which use community-based treatment programs, often fail to keep all of the physiological and social aspects of addiction in view. As a result, they can sometimes exacerbate the problem. When we isolate people or incarcerate them, we make it more likely that they will seek out other people who are addicted for social support. This is nothing more or less than the situationally rational attempt to preserve a degree of control or autonomy. But this leads to the continuation of the cycle of addiction. So, we must make a choice: our collective actions can provide support to those who struggle with addiction, or they can provide barriers to recovery. We are all entangled with the human vulnerability to addiction, and there is no way of withdrawing from the problems and hoping that they will just go away.

Aristotle and many other thinkers have emphasized forming and sustaining habits as key capacities related to self-regulation. Practices of child rearing and education attempt to shape habits in plausible ways. In the best cases, they help people learn to avoid situations where they might be in danger of making bad decisions.[58] But we are biologically prepared to use opioids, and we are vulnerable to their effects—good, bad, ugly, and deadly. Where the use of opioids ameliorates pain that impedes well-being, using them can be adaptive; but where they become linked to the avoidance of pain and the continual search for relief, this can overwhelm the desire for well-being and carry us into addiction. The adaptive response is far too similar to the nonadaptive response that compromises our options. Addiction narrows our focus to short-term gratification, and as addiction progresses, it becomes a predominantly avoidant process, which is more focused on escaping withdrawal than achieving a high. When this occurs, addiction *becomes* a fundamental need without which the addict feels their interests are threatened. Dopamine is involved in all of this.

Just as importantly, addiction is always *both* social and individual. It involves the larger community and the kinds of resources that are available for navigating the effects of addiction. In many cases, beating addiction requires having others who care and are supportive. Sometimes, this can be as little as the inspirational figures who show us other ways of being in the world. But the pain of addiction is isolating, and the solution requires breaking that isolation. The desire to punish drug use and addiction is a blunt instrument that often ignores the cultivation of the tools that are necessary for people to live their lives to the fullest and achieve higher degrees of self-control.[59] As a society, we need to find better ways of giving people the resources they need to avoid or minimize vulnerability to addiction. Of course, we live in a world pervaded by unequal treatment, unequal access to resources, and cultural institutions that harm specific classes of people. The playing field is not level; it never has been, and this will always matter. We don't know exactly who is vulnerable because there are complex interactions between physiological and social contributions to addiction. But we know people need better options, especially during adolescence when habits are still developing and people are more vulnerable to the entrenchment of habits that are bent toward addiction. Regardless of age, however, many people may simply be predisposed to addiction by genetics, history, family structure, or social pressure.

In many contexts, however, social contact and social support will be critical to breaking the cycle of addiction before it destroys lives. Childhood trauma and experiences of being devalued will often be part of the addictive trap, as will feelings of being devalued, which can lead people to pursue avoidance strategies while decreasing opportunities for seeking greater meaning or building tighter connections with others.[60] Finally, escaping addiction will almost always require imagining a life beyond addiction. Sometimes, this will be a matter of seeing options that others pursue or observing the effects of better decisions. Sometimes, it will require mimicking the path that someone else has taken. But often, it will require building relationships with others that scaffold the construction of better habits, provide support where necessary, and support the practices of praising and blaming that make it possible to follow a new path. If we understand freedom and autonomy to require constructing habits that will allow us to pursue our long-term interests, then we can immediately understand the importance of cultivating patience and perseverance.[61] We want to learn to avoid what is harmful, and being free requires the patience to avoid what is most immediately pleasant as well as taking the time to develop a better understanding of our own motivations and limitations. Addiction, because it consumes attention, makes this process difficult if not impossible.

2

CRAVINGS AND
OTHER MOTIVATIONS

I n chapters 2 and 3, we offer a more detailed exploration of some of the diverse forms of expectation that organize action in typical and addictive contexts. A critical part of this story derives from the neuroscientific discovery that the brain and body produce endogenous opioids, which underly a range of adaptive and regulatory behaviors. The endorphins are a specific class of neuropeptides, and peptides and neuropeptides are short chains of amino acids. These chains have been evolving for nearly a billion years, and the same molecules can be found in bacteria, plants, flies, fish, and mammals.[1] In each case, these peptides and neuropeptides play numerous roles in supporting the regulation of the internal milieu. In part, this is possible because evolution can operate through gene duplication and diversification, allowing the same peptide to play multiple roles throughout the brain and body.[2] Where this happens, it yields a form of speciation-through-diversification: genes are duplicated; the proteins and processes they produce are put to different uses; and over time, this yields complex chemical signaling systems that make use of various peptides and neuropeptides.

The first endorphin to be characterized in the 1970s was β-endorphin, a 31-amino acid peptide; a-dynorphin, a

17-amino acid peptide was sequenced soon thereafter.[3] Each of these endorphins has an affinity for a specific opioid receptor: β-endorphin has an affinity for the kappa-opioid receptor; a-dynorphin has an affinity for the mu-opioid receptor. Candace Pert and Solomon Snyder carried out the initial research that elucidated the structure of the opioid receptors.[4] Prior to this discovery, it was assumed that such receptors might exist, but they had never been isolated. Pert was a graduate student at Johns Hopkins University in the 1970s, and she was one of the neuroscientists who was responsible for discovering the brain's pain relief system. Unlike her male colleagues, however, she was passed over for the prestigious Lasker Prize, which is often a precursor to the Nobel Prize.[5] Of course, this wasn't the first time that the contributions of a graduate student who was instrumental in making a scientific discovery have been ignored, nor is it likely to be the last. But it was significant that Pert was the only woman in this group, and the abject unfairness of this situation yielded lasting pain.[6]

Science, like all other human activities, can be competitive and unfair. But giving credit where it is due tends to further scientific collaboration, in many cases, far more than cutthroat competition. Our current understanding of the brain's endogenous opioid system reflects the contributions of diverse scientists throughout the world, and it has only been possible because of a massive, cooperative exploration of various biological processes. Over the last fifty years, research in numerous labs using diverse empirical techniques has uncovered four main types of opioid receptors: delta-, kappa-, and mu- receptors, and a nociception/orphanin receptor.[7] The binding conditions for various endorphins have been measured using immunohistochemistry, and RNA levels have been measured using radioactive probes.[8] The specificity of various peptides has been measured under typical conditions and in the context of blockers using electrophysiological measures. Animal models have been used to enhance or eliminate the availability of neuropeptides, and recently, optogenetic techniques have been used to enhance or eliminate the availability of endorphins.[9] Collectively, these approaches have yielded a range of insights into the functions of endogenous opioids, and they have revealed the presence of opioid receptors across numerous brain regions.[10]

In this chapter, we examine some of the roles played by these systems. We argue that the same peptides, neuropeptides, and receptors are implicated in the psychobiology of craving in any context where locally relevant challenges and opportunities must be addressed and managed. This matters because it means that the same systems are implicated in contexts that are essential to adaptive regulation and contexts that are involved in the forms of dysregulation indicative of addiction.

RECEPTORS, NEUROPEPTIDES, AND GENE REGULATION

Opioid receptors have been shown to be present in all vertebrates, with an origin that can be traced to half a billion years ago.[11] They have also been found in some invertebrates, which lack many other features of vertebrate neural design.[12] It should come as no surprise that endorphins transmit a diverse class of messages and that we are only beginning to understand how they affect neural cells when they are produced endogenously as opposed to being introduced into the brain as narcotics.

Across species, peptides are entangled with broader networks of chemical signaling systems, which collectively sustain adaptive regulation by shaping intercellular communication and transforming responses to the broader ecological context. To affect neural activity, most neuropeptides must be secreted into the brain's extracellular fluid. Sometimes, they are sent from diverse regions of the brain and body, and at other times, they are produced and used locally, but they always function by binding to neural cells and shifting patterns of cellular activity. Some peptides, such as vasopressin, oxytocin, and angiotensin, enter the brain through regions that are not protected by the so-called blood-brain barrier. But the endogenous opioids are produced and used in the brain itself. This is intriguing because the highest concentration of opioid receptors seems to occur in regions that are implicated in regulating mood and managing pain.[13] Indeed, the brain seems to produce substances like morphine and codeine to regulate diverse experiences of pain and discomfort.[14] Perhaps unsurprisingly, the opioid receptors also appear to traverse many brain regions, with a focus on regions that are involved in forms of hedonic assessment.

The endorphins interact with a range of neurotransmitters, including serotonin, dopamine, norepinephrine, GABA, and glutamate. These neurotransmitters shape numerous patterns of thought and behavior. They also regulate the search for resources, modulate incentive salience and risk tolerance, and organize attentional profiles. The same chemical signaling systems that are implicated in the search for nutrients and other life-sustaining resources are also involved in the search for drugs, which forms a crucial part of opioid addiction. As we argued in chapter 1, dopamine is involved in mobilizing and organizing action and keeping track of the salience of diverse incentives—in addictive as well as nonaddictive contexts.[15] Likewise, vasopressin, oxytocin, and corticotrophin-releasing hormone are all involved in regulating bodily needs and producing and organizing emotions. Here, too, addiction bends these forms of adaptive regulation in nonadaptive ways, and it does so—at least in part—through the interactions between endorphins and other chemical signaling systems.

All of this matters because it helps to make it clear that addiction is not simply a matter of opioid dependence. The brain and body come prepared to manage various forms of pain and adversity, and the endorphins are tied to both the management of pain and the experience of euphoria. The release of endogenous opioids in aversive contexts makes it easier to tolerate discomfort. In many contexts, these neuropeptides are linked to regulating inflammatory signaling molecules. These neuropeptides also support vulnerability to addiction because they are able to produce a euphoric sensibility and because they interact with networks of "information molecules" that shape cravings, attention, and memory. But in each of these cases, the opioids and the opioid receptors appear to play a critical role in modulating these diverse systems.

Neuropeptides are regulated in development and throughout the life cycle through processes that are under epigenetic control. Like all biological products, neuropeptides are produced through DNA/RNA interactions, and the epigenetic regulation of endogenous opioids, in response to events that impact the brain and body, is one of the key mechanisms by which bodies adapt to their circumstances to maintain viability.[16] The idea is that genes can be silenced or expressed without impacting large-scale genetic structures. Variations in gene expression occur in a range of contexts and, in many cases, experience changes how genes are expressed. The genes that are involved in the production of endogenous opioids change their expression over the course of days, weeks, months, and more in ways that depend on patterns of changing circumstances.

This process has limits in the context of increasing adaptive responses and increasing vulnerability toward addiction.

Importantly, there is evidence that changes that occur in the context of adversity impact the likelihood and severity of opioid addiction. The expression of peptides can increase or decrease in different parts of the brain and body as the distribution of challenges and opportunities changes. It has long been clear, for example, that endorphins have variable conditions of expression and efficacy, which depend on both the context and history of their use.[17] For example, repeated expression of opioid peptides often provokes dysphoria, as well as feelings of unhappiness or uneasiness.[18] Moreover, diverse conditions of experience impact both the expression and regulation of endogenous opioids in ways that appear to be tied to the vulnerability to addiction in some individuals.[19] In animal models, studies have shown that early life experiences can create a vulnerability for opioid addiction by impacting the structure of brain chemical signaling systems and by modulating the salience of events that are tied to drug use.[20] This results in a behavioral phenotype that is more resistant to the extinction of learned associations and increased tendencies to relapse. What does this mean? We can get a sense of why this is important by thinking about animals who will continue to press a bar for access to opioids long after access to them has been revoked. In contexts like this, the goal and the incentive persist even when the reward is absent.

In an attempt to grapple with these issues in the context of opioid addiction, a great deal of recent research has focused on dynorphin peptide expression and the role of the mu-receptor.[21] The mu-receptor seems to play a critical role in disassociating pain relief from euphoric sensibilities, and chemicals have been developed that target mu/delta or mu/kappa receptors more selectively to affect these kinds of sensibility in different ways.[22] This research has yielded substantial and important results—though the normative status of the results is rarely straightforward. For example, both methadone and fentanyl are selective mu-receptor agonists. By contrast, substances like nalorphine and naloxone are nonselective antagonists, which block the activation of opioid receptors more generally.[23] In the context of basic research, these chemicals have played important roles in fostering an understanding of the neural mechanisms of addiction. Far more importantly, they have also provided a way to reverse overdoses.[24] Unfortunately, they have also yielded a far more deadly opioid: fentanyl.

Alongside these kinds of research, experimentalists have also started to tease out the specific DNA sequences that underlie the impact of opioid receptors in the context of pain and dependence. These encoding mechanisms have been elucidated by laboratories across the world.[25] Moreover, increasing knowledge about the molecular structure of these molecules has made it possible to study physiological genetics at both the receptor and the neuropeptide levels.[26] It has become clear, for example, that expression levels change in response to evolving circumstances and that these changes can be reversed. One candidate gene, OPRMI—and varying promoters (for example, exon 1 and exon 2)—appears to encode important information about the mu opioid receptor, which is essential for the analgesic impact of the opioid-related substances.[27] Systematic meta-analytic reviews of the literature tie this gene to opioid dependence.[28] But the impact of this gene is vast. It has an influence on respiratory depression, the inhibition of gastrointestinal function, and variations in neurotransmitter expression—for example, changes in the release of dopamine that leads to the neglect of hunger and other bodily needs as well as opioid dependence.

Finally, it has become clear that mu-receptor activation often has inhibitory functions mediated by G protein activation.[29] This is only one part of the story. The expression of endogenous opioids is regulated by various steroids—such as estrogen, testosterone, and cortisol—all of which are vital constituents of the drive to forage for food, shelter, sex, and social contact. Likewise, interactions with peptides that organize affective and attentional profiles can shape the likelihood of addiction.

Consider, for example, the elevation of corticotrophin-releasing hormone in aversive contexts.[30] This neuropeptide is commonly linked to the experience of fear, despair, and depression.[31] It is also known to trigger the release of dynorphin and be colocalized with GABA.[32] The overexpression of neuropeptides like corticotrophin-releasing hormone can result in increased vulnerability to trauma and higher rates of self-medication later in life. Early experiences of trauma, even transient ones, can cause lasting despair-like behaviors that are regulated by this molecule in diverse animal models.[33] The main thing that we want to highlight with this discussion is that there is no clear or obvious pathway from a story about evolution, genetics, or development to a story about the nature and structure of addiction. There are numerous interacting factors in play, and any plausible story must draw upon all of these factors and highlight the ways that historical and ecological constraints shape addictive profiles. That said, there are many things that we have learned about the impact of biological constraints on the experience of addiction. In the remainder of this chapter, we turn to some of these factors.

CRAVING

Mobile animals rely upon sensory systems to interact with the world in contexts that include appetitive search, ingesting food, and seeking out the satisfaction of needs. Searching for and exploiting nutritional resources is a continuous and active process.[34] Sensory systems also guide searches, especially when interoceptive systems anticipate or detect changes in our needs that demand behavioral and physiological responses. We probe the world for events that are related to nutritional resources as well as potential threats.[35]

Gustatory systems have ancient origins, with taste and olfaction predating the emergence of vision and hearing. Gustatory systems are found both in invertebrates and vertebrates. While there is variation around feeding patterns, these sensory capacities underlie the search for the nutrients and minerals that are needed to keep the body running. When we are hungry, taste and olfaction lead our search for what is needed. The hungry fly and the foraging bee search for food with chemo-sensory systems that are not so different from our own.[36] In our own case, these sensory systems are tightly coupled to the hedonic systems that inform us about the likelihood that something we eat will kill us, regardless of whether we liked eating a specific food or not. So, we readily learn what to avoid when we get sick following the ingestion of a food source.[37] We readily learn to track the foul smells and bad tastes that are associated with some dangerous foods, and we readily learn to approach foods that smell inviting or taste good, especially where they are potentially nutritious.[38] It is clear that food selection is helped by these core properties, which guide choice, ingestion, and digestion.

We have all experienced cravings, but they are more difficult to describe from a biological perspective. A craving is an intense and urgent desire to consume a substance. The craving for sweets for example, can sometimes signal a need for energy. In our current ecological context, however, many sweets have little value beyond the satisfaction of cravings. The allure of sweets remains, but in a world dominated by sedentary living, this might sometimes result in higher levels of metabolic diseases. That said, despite widespread access to sweet tastes, many people navigate our culture without becoming entangled with excessive cravings for sweets. This is intriguing. For others, the allure of sweets is mindboggling, and the endless availability of sweet products has a range of negative impacts on many people.[39] Just as importantly, most people who do have a sweet tooth are not addicted. Addictive cravings are always pathological in the sense that they push beyond what is adaptively necessary, and sometimes, these pathologies emerge in contexts like seeking food, sex, gambling, or danger, which might be adaptively relevant.

In many cases, cravings are directed toward particular nutrients, and cascades of internal processes are initiated alongside cravings to prepare the body to use necessary resources, such as carbohydrates, protein, fats, and micronutrients. This is clearest in the context of cravings, which signal what we need almost directly.[40] Consider a craving all of us have probably experienced at some point: thirst. The craving for water is easy to elicit because the ingestion of water is essential for maintaining fluid balance in our bodies. It can be triggered by many different things, including eating salty foods or losing blood.[41] Many animals want to ingest water immediately after consuming salt since they need both to regulate extracellular fluid volume.[42] Hormones like angiotensin and aldosterone are motivated to support the conservation and ingestion of salt as well as the craving to find sources of salt when the body senses a need.[43] Alan Epstein claimed that excessive salt ingestion was an addiction—but this claim stretches the definition of addiction beyond its breaking point.[44] That said, sodium hunger is a biologically basic appetite. Therefore, considering sodium hunger can help to shed light on the nature of craving.

To begin with, many mammalian herbivores must search for minerals to ingest, and some large herbivores, such as elephants, will travel incredibly long distances to ingest salt. Strikingly, the taste of salt activates specific sodium detectors in these and other species.[45] But diverse regions of the brain are employed to remember *where* sodium has been found, regardless of whether it is needed; *how* sodium has been acquired; *what* associations there are between sodium and other tastes; and *when* sodium has been available—over timescales ranging from the time of the day to a specific season.[46] The craving for sodium is widely expressed across a wide array of phyla, including primates.[47] But this craving is shaped by current regulatory demands and histories of past experience. Many animals find seawater (3 percent salt) repulsive and aversive. But seawater can become attractive if an animal is hungry for sodium, leading salt-depleted rats to both ingest sodium and display facial expressions associated with something palatable and tasty, like sucrose.[48] Strikingly, the craving for salt is immediate in the context of salt deprivation—even for rats who have *always* previously displayed an aversion to seawater.

Studies have long shown that animals also select food on the basis of specific cravings for carbohydrates, fats, and other lipid-related nutritional resources.[49] Curt Richter, for example, explored the behavioral regulation of nutrients, fluids, and other resources.[50] This research, as well as subsequent research in a similar vein, has shown that behavioral regulation always occurs in concert with the physiological regulation of the internal milieu. This is important, as regulation often reflects experiences of past resource deprivation, especially where there is insufficient access to sodium, water, or other life-sustaining requirements. Moreover, it reflects changes that occur across various stages of life, as the need for different resources can shift as a result of development, pregnancy and lactation, normal activity, and aging. In each of these contexts, numerous peptides and steroid hormones support abilities to track and respond to physiological needs, and all of these peptides and hormones are calibrated by diverse anticipations. These molecules convey messages from diverse bodily regions and do so in ways that reflect patterns of past actions and differences in biological starting points.

One well-known hormone involved in metabolic regulation is insulin, a peptide that is produced in the pancreas to regulate glucose levels.[51] Ivan Pavlov pointed to the role of anticipated needs in regulating changes to the internal milieu to prepare for the digestion of food.[52] But differences in ecological contexts and physiology can affect the way that this system is calibrated.[53] A wide range of peptides that are produced in the stomach and the small intestine—including bombesin, CCK, insulin, leptin, and neuropeptide Y—play further critical roles in shaping the craving for food, utilization of nutritional resources, and the experience of satiation after we eat.[54] Moreover, all of these chemical signals are calibrated by diverse regulatory demands.

In general, cravings alert organisms to pressing needs in ways that are sensitive to fluctuations in the internal milieu, including unexpected changes in what pursuing viability demands. But just as strikingly, cravings can also go awry in ways that can cause people to pursue harmful or useless substances. Exaggerated appetites can sometimes become addictions, and typical members of our species are naturally predisposed to abuse some substances. For example, some people crave sources of starch and sweet-tasting foods that are high in calories but low in nutritional value. The sweet tooth is a natural predilection that serves the needs of many animals. In many ecological contexts, sweet tastes play a critical role in the search for sources of metabolic energy.[55] This is because sugar often serves as a signal that a food—maybe a fruit—contains high-quality nutrients that are easy to extract. In other ecological contexts, however, the craving for sweet-tasting foods can become a vulnerability. We manufacture substances that increase sweetness independent of increases in nutritional value, but people track factors that satisfy their cravings even when the broader value of a signal changes.

Another way to put this point is that appetites can *look like* addiction, especially where they result in tissue damage, loss of capability, or the narrowing of focus. But the mere presence of such an appetite does not entail that someone is addicted to the thing they crave. Consider a natural but confusing and disordered craving known as pica. Pica can occur under many conditions, including pregnancy. It is a nonspecific craving for something that is not nutritional. It often leads people to eat chalk, for example, which is not nutritious, and it can often culminate in the ingestion of things that are harmful. In this context, a *craving* seems to emerge for something that is neither nutritious nor necessary. At a high level of generality, this is also true of opioids, though we should remember that they do have a deeply important capacity to help people navigate pain and trauma. Opioids are not necessary for health or flourishing, but they can open pathways to health and the minimization of discomfort. In some cases, they can even support flourishing in the long run. But just as importantly, they are distinctive in the ways that they can distort evolved tendencies to seek out natural rewards and manage various emotions—including social emotions.[56]

What makes them distinctive? In part, it is the fact that endogenous opioids are often recruited in the context of regulating bodily needs.[57] They are also implicated in the biology of pleasure and the hedonic aspects of decision making. This has led some people to assume that opioid users must be pleasure seekers.[58] But this is an overly simplistic perspective. Many recreational users report nausea or physical unease after the initial use of opioids, not pleasure or euphoria. Some of these people come to crave these substances, but it would be a mistake to assume that pleasure seeking is always the craving that organizes opioid use and abuse. Some people turn to narcotics simply for the novelty of the experience. This is where facts about adolescent risk-taking are likely to matter most.[59] But far more commonly, the craving for opioids is motivated by a search for relief from physical or psychological pain.

The VA Center for Addiction Studies emerged in the 1970s to respond to the needs of veterans who were returning from the Vietnam War.[60] Research in this context made it clear that opioid craving was not simply a hunger or a drive to pursue pleasure. It reflected a bodily demand and an acquired need. Hunger, thirst, and the craving for necessary minerals are deeply rooted in physiological demands, and they direct us toward the restoration of bodily tissue. By contrast, the craving for opioids is a response to the modulation of chemical systems in the brain. This craving changes over time, and it can become more severe after longer periods of use. However, there is no direct link between the craving and a needed resource. Opioids are often sought to minimize physical and psychological pain. But for these veterans, cravings were rarely driven by the experience of discomfort or withdrawal, and the minimization of pain did not always appear to be essential to the craving. Instead, there was an anticipated need, decoupled from immediate discomfort but anchored to memories and to the anticipation of impending discomfort.

HEDONIC SENSIBILITY

Philosophers across diverse traditions have sought to explain why we seek pleasure. We find different answers in the works of

Aristotle, Bentham, Candrakīrti, Epicurus, Mill, Śāntideva, and the Stoics. But in each case, it is acknowledged that we attempt to maximize the experience of pleasure and minimize the experience of pain.[61] We are drawn toward pleasant or rewarding events, and we tend to avoid noxious or painful ones. These tendencies to approach and avoid have been an important driver of animal evolution. Therefore, the hedonic experiences that are orchestrated by the brain play a prominent role among the competing needs and interests that animals must navigate.[62] That said, decisions to maximize pleasure and minimize exposure to noxious phenomena tend to operate on the basis of rules of thumb, which evolution has entrenched because they tend to keep animals alive. As we explore the world, we associate memories with good or bad feelings—and the resulting emotions are both reflections of our response to the world and the phenomena that shape how we experience the world. But importantly, our experiences of pleasure and pain are not always attuned to our best interests. Some of us take pleasure in eating foods we know are bad for us. Some of us enjoy listening to music at volumes we know will cause hearing damage. And hedonic urges can lead us to pursue things we "want," even when we know there are better alternatives.

As we have already noted, many animals do not find their first encounter with opioids pleasurable. But the addictive quality of opioids can come to outweigh initial negative experiences. This is a quality that opioids share with many other substances that can become "acquired tastes," including things like quinine and chili peppers in rats.[63] For a typical rat, by contrast, the taste of morphine elicits a mix of positive and negative responses, much like the experience of ingesting hypertonic sodium.[64] However, when rats are exposed to morphine over the course of development, they tend to like how it tastes, unlike rats who have only been exposed to pleasant tastes or rats who have previously been exposed to quinine. The aversive response to sodium is not fixed, and it changes with need and exposure, yielding shifts in both incentive salience and hedonic value.[65] For example, when sodium is needed, seawater is perceived as pleasant—and this can happen within seconds as the body prepares to face this challenge. By contrast, continued and repeated exposure is necessary for opioid addiction to take hold. In each case, though, endogenous opioids underlie adaptive responses to changes in bodily needs. This is true in the context of sodium hunger, but it is also true in the case of coming to like the taste of morphine.

In each of these cases, regions of the brain are recruited, including central and medial regions of the amygdala.[66] Mu-opioid receptors in these regions play a critical role in transforming hedonic responses to salty tastes.[67] The parabrachial region of the brainstem is also involved in sodium intake.[68] Here, too, there is a strong influence of endogenous opioids.[69] These regions are densely, bilaterally, and directly connected to one another, and they are some of the critical sources of opioid-driven changes to gustatory sensibilities, which rapidly arise in the context of sodium hunger.[70] Small amounts of morphine can also enhance responses to quinine and high levels of salt.[71] This change is supported by the activation of endogenous opioid receptors.[72] Finally, pharmacological blockades of endorphins disrupt the liking of diverse substances, including morphine.[73]

Such data suggest that endorphins can transform aversive substances into ones that are liked. But just as importantly, these systems also allow animals to acquire preferences for things that are associated with the satisfaction of cravings. In some contexts, this can yield patterns of liking and ingesting that are only contingently associated with a need (for example, craving potato chips as a result of their association with sodium). In this context, substances are tagged as rewarding because they resemble past associations. The brain comes to predict a reward when a taste memory is triggered, and this incentivizes the search for objects that do not have value to us intrinsically.[74] In some cases, this can even cause appetitive responses to tastes when a resource is not needed.[75]

Across each of these contexts, endogenous opioids play numerous roles in the experience of basic likes and dislikes. They also underly changes in attitudes toward needed resources that might otherwise be unpalatable. But there is almost no end to the range of positive feelings that can be associated with endogenous opioids. Consider the runner's high, which is an incentive for many people to remain active in our more sedentary culture. Endorphins are at play both in anticipating the run and in the experience of running.[76] The feelings of anticipation and the experience of well-being would be impossible without the neurotransmitters and neuropeptides that organize action, prioritize strategies, and support feelings of satisfaction. Endogenous opioids also underlie the sense of sweetness, and they enhance the hedonic impact of sweet and other rewarding or good-tasting substances.[77] They are also implicated in every context where things become alluring to us. Finally, they help to minimize forms of physical and psychological pain. This is the adaptive side of endogenous opioids. They enhance consummatory behavior, and they make life—and the things that are needed to sustain it—pleasurable. However, the same molecules are implicated in the adaptive regulation of the internal milieu and in the dark roads that lead to addiction and behaviors of self-harm.[78]

The problem of addiction is that a person experiences the ingestion of opioids differently; they see reminders of a high in the environments where they are embedded, and this allows the brain to be reawakened to craving mostly without conscious thought. This is why relapse is prevalent and heartbreaking. Addiction is not adaptive, even though the amelioration of pain and the ability to find resources to do so is. Opioids can cause short-lived experiences of well-being, but negative emotions characterize many individuals who experience addiction. This is especially true of people who experience the pain of withdrawal as well as the psychological effects of addiction. But this is a mechanism by which opioid addiction feeds itself: addiction increases negative affect, making the high more necessary to the user.[79]

ANTICIPATING DANGER

The anticipation of danger and pain is also a fundamental principle of neural design.[80] Many animals possess rich capacities for avoiding and managing pain. Despite the simplicity of their brains, even flies must anticipate where they are most likely to find resources and where they are most likely to encounter danger. Many animals also display protective drives and strategies for avoiding further harm. Finally, nociceptors have been found in a range of animals, including insects, leeches, crustaceans, mollusks, reptiles, fish, and mammals. It seems likely that sensitivity to pain is an evolved solution to the challenges of tracking danger, being motivated to protect oneself from harm, and avoiding situations where damage is likely.[81] It should come as no surprise, then, that sensitivity to pain is highly heritable. But sensitivities to different pains are only weakly correlated, so someone who is highly sensitive to pain caused by heat might, for example, be less sensitive to pain caused by pressure.[82] This suggests that whatever the evolved capacities for pain are, they are likely to be highly complex. Indeed, it might be a mistake to assume that there is a single kind of process involved in every experience of pain.

This situation is further complicated by two facts. First, a substantial portion of the human brain is involved in assessing and responding to various forms of pain (figure 2.1). Experiences of pain are regulated by complex interactions between the peripheral and central nervous systems and the corticospinal pathways.[83] And it has long been clear that responses to painful stimuli are shaped in profound ways by brain-based mechanisms and not just by signals of damage from peripheral receptors.[84] Second, bodily and psychological trauma can both evoke complex cascades of neural activity, which ripple throughout the central nervous system, to bring the demand to protect ourselves

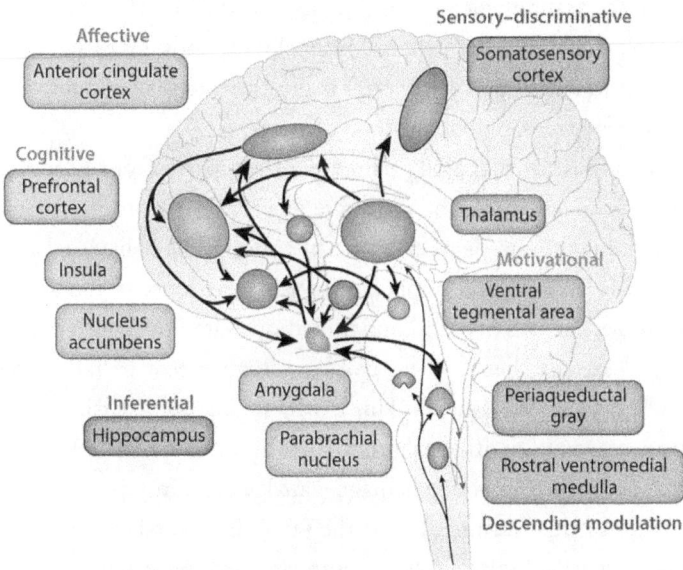

FIGURE 2.1 A large interconnected neural network of brain circuits transforms nociceptive information ascending from the spinal cord into an aversive, painful experience. Reprinted with permission from Corder et al. 2018, 461, figure 3a.

to center stage.[85] For this to happen, numerous neuropeptides are mobilized, including corticotropin-releasing hormone and endorphins, to regulate behavioral responses to painful stimuli.[86] Sometimes, endorphins are even mobilized preemptively to prepare for potential bodily damage and to prepare the brain and body to navigate challenges that must be faced.[87]

The complexity of this system is one of the primary reasons why pain cannot be simple or unified. Some pains are riveting. Others are overwhelming. Some have a slow and deadening rhythm. Others spark a reaction before we feel them—as when we notice a burn only after we pull our hand away from a hot stove. Additionally, toothaches feel different from stomach aches and aching joints, and these physical pains differ substantially from the psychological pains that pervade everyday life for many people, including the pain of losing close friends, excessive shyness, and failure and rejection. Finally, the intensity of pain varies across times, individuals, and contexts.

Abilities for managing pain are a deep part of our adaptive behavioral, affective, and cognitive repertoire. According to one recent hypothesis, pain serves as a warning.[88] The behavioral and physiological changes implicated in pain experiences—including increased sensitivity and inflammation—are inexpensive and useful signals that help us avoid the higher cost of failing to address potential dangers. This is why the reduced sensitivity to pain associated with analgesics sometimes supports behaviors that cause further bodily damage—analgesics minimize our sensitivity to this warning. But at the same time, persisting experiences of pain limit our capacity to act, yielding experiences of depression and the diminishment of other motivations—even though pain "is well adapted to make a creature guard itself against any great or sudden evil."[89] Whether pain signals a risk of bodily damage or problematic aspects of social isolation and degradation, it tends to suggest that something is wrong, and it tends to motivate patterns of thought and action that will allow us to respond to these challenges. Here, too, the story is complex.

In some cases, encounters with noxious stimuli will trigger activity in nociceptive neurons, evoking the release of neurotransmitters, such as glutamate, to propagate nociceptive signals throughout the widely distributed pain systems. This process can be modulated by the release of endogenous opioids or by the ingestion of exogenous opioids. In some contexts, endorphins are even secreted to inhibit pain where further action is necessary.[90] The ability to flexibly manage experiences of pain might be just as important from a biological perspective as the ability to feel pain in response to the threat of further damage. In most cases, feelings of discomfort and increasing sensitization are prominent features of pain—and this is a good thing. A heightened sensitivity to pain reduces the use of damaged tissues, minimizing movement and giving the body an opportunity to heal. In typical contexts, pain tends to diminish as the threat of further damage dissipates.[91] But critically, processes for combating pain can begin to degrade neural tissue if they remain active for too long, as they do in the context of chronic opioid use. Chronic opioid use also reduces their effectiveness for managing emotional pain—and in many contexts, this actually ends up increasing emotional pain: on the one hand, withdrawal increases the sensitivity to pain; on the other hand, it also increases the motivation to relapse to avoid this experience of pain.[92]

The problem here is familiar and not simply a matter of addictive pathology or neural dysregulation. Pain is often difficult to control. You can try to treat the underlying causes, you can try to learn to live with it, or you can try to ignore it. But since pain is a warning of a danger that must be addressed, it is hard to ignore. Persisting physical or psychological pains, or pains that are difficult to manage in other ways, often lead people to opioids. This makes sense, as opioids can make life bearable—at least for a short time. To the extent that they ameliorate pain and discomfort, they can play an adaptive role in human life, and the medical benefits of opioids begin to look like one of nature's greatest gifts.[93] But pursuing and consuming opioids can easily become pathological, as the drive to mitigate pain becomes entangled with the use of opioids. The exaggerated expression of endogenous endorphins decreases pain sensitivity for a time, but the continual elevation of opioid activity in addiction yields a heightened sensitivity to pain.[94] Heightened pain sensitivity is also observed when people are treated for opioid addiction with buprenorphine, an opioid receptor agonist.[95] It does not matter if the initial pain was physical or psychological. Over time, both psychological and physical pain can exacerbate the vulnerability for relapse. Sometimes, people take opioids for pain control. Other times, they do so because they inhabit a cultural context where opioids are used to escape an otherwise awful situation. But once the relief from pain becomes necessary for mental well-being, the possibility of addiction rapidly comes to the fore.

The experience of pain often serves as a signal that something is wrong, and it often leads us to avoid actions that will cause further harm.[96] Sometimes, this signal can persist beyond the point where it is helpful, so doctors prescribe analgesics to numb the pain while they treat underlying causes. But numbing pain can be dangerous, especially in the case of psychological pain. Opioids deaden emotion and quiet emotional discomfort. But outside of the experience of the high, many sources of psychological pain persist as a result of systemic harms and systemic marginalization. Often, the life changes that are necessary to treat psychological pain are not materially possible, but people want relief, so this situation presents a real vulnerability to addiction—and an escape into narcotics is often the most available option.[97] But the effects of chronic opioid use are systemic. They shape the activity of the frontal cortex in ways that impact diverse cognitive capacities, including capacities for social- and self-regulation. The frontal cortex is integrated with regulatory systems throughout the brain—including the nucleus accumbens, bed nucleus of the stria terminalis, and the amygdala—and all of these systems are implicated in endogenous opioid activation—in both adaptive and pathological contexts. Opioids remain a part of our medical arsenal because they are effective in treating pain that other medicines cannot treat. But because of these large-scale and systemic effects, their use can also skew perception, emotional experience, and decision making in profound ways.

In the context of addiction, numerous features of experience are bent toward the pursuit of a specific substance or activity, such as the use and abuse of opioids.[98] Importantly, addiction does not happen in a vacuum. A person becomes addicted to opioids within a specific social cultural context, where they are a member of specific social groups, some of which have specific histories of advantages and disadvantages. The significance of this inherently social context is inescapable, but easy to forget.[99] Most people who are exposed to opioids, even those who become dependent upon them, do not become addicted.[100] However, dysfunctional environments, unfair and discriminatory environments, and fearful and abusive contexts lend themselves more readily to abuse.[101] Moreover, as the historian Claire Clark notes, "Whether its source is the brain or society [addiction] is essentially a form of suffering."[102] Finally, the experience of addiction shapes the ways that people resonate and interact with one another, and it shapes the likelihood that they will engage in self-destructive acts together. In light of all of these complexities, addiction should be understood as a form of anticipatory and cephalic dysregulation, which occurs in a specific context— where resources are available or lacking and against the backdrop of specific needs and expectations.

The opium plant, pills, or heroin can all come to serve as objects of desire in ways that share much in common with the salt lick for herbivores. But sodium-hungry animals satisfy a number of mineral needs when they visit a salt lick, and this serves true biological needs.[103] Moreover, in most cases, when sodium returns to a viable level, the craving for salt will rapidly fade. The craving for opioids can feel like agitation, and scoring drugs can yield relief or excitement, or it can placate tension and discomfort. But the pursuit and acquisition of drugs can also become a driver of addiction. When the high is achieved, it will often be less intense than when the drugs were first used, and this can lead to the acceleration of the cycle of pursuing and consuming opioids.

The dangers of morphine were noted in the middle of the nineteenth century. Early in the twentieth century, T. D. Crothers asserted that "morphinism is a modern disease and threatens to be one of the most serious menaces accompanying the 20th century."[104] He worried about "needle mania," and described the many faces of morphine addiction and withdrawal, including sleep disorders and depression. He called attention to comorbid drug use, including the tendency to switch from alcohol to morphine. Recent research suggests a reason for this tendency: alcohol and other drugs often trigger the release of endogenous opioids and peptides like corticotrophin-releasing hormone.[105] Crothers also described the way that culture could influence consumption, suggesting that Thomas De Quincey's treatise about the positive allure of opium use had led innumerable people to seek but fail to find "the same results from the use of opium."[106] Many other investigators, including Benjamin Rush and Thomas Trotter in the eighteenth century and Edward Levinstein in the nineteenth century, argued that addiction was a brain disease that reflected an uncontrollable desire to use an addictive substance. This position has also been accepted by the American Society of Addiction Medicine, which describes addiction as a chronic disease of brain reward.

However, we should tread cautiously. At present, many aspects of neuroscience remain poorly understood, but they won't always be—and our knowledge is constantly increasing. Moreover, we do know that addiction takes over the whole body. Addiction shapes what is recognized as a need. Addiction shapes habits. Addiction shapes what a person craves and what they pursue. In each of these cases, patterns arise through the same features of neural design that maximize our predictive capabilities.[107] When a person is exposed to the allure of opioids, they can become caught in an intolerable bind: more drugs are consumed but less and less pleasure is experienced, and this can yield stronger and stronger cravings. At this point, the effects of addiction can pervade a person's life, shaping everything from waking up in the morning to whatever sleep means for the person. A diverse set of neuropeptides and neurotransmitters are involved in all of these facets of opioid addiction, affecting the way that people chase the high, enjoy it (or not), and withdraw from it.

There is no doubt that capacities for tracking and responding to rewards are impacted in this process. Indeed, addiction can be seen, in part, as a form of cephalic and neural dysregulation where the pursuit of rewards and other incentives becomes an obstacle to an individual's ability to thrive and flourish. However, the centrality of endorphins to our well-being makes it hard for some people to refrain from abusing drugs that increase their prevalence. The craving for opioids reflects the way they engage a broad array of endogenous chemical signaling systems, all of which are involved in the navigation of physiological, ecological, and social challenges. Many regions of the brain are recruited during opioid use and abuse, and the range of changes in the brain are mindboggling. For this reason, it should come as no surprise that there is no *one* form of craving that is distinctive of addiction. There are distinctive patterns of behavior, and these common patterns of use and abuse often lead to physical dependency, overdose, and death.[108] But we should always remember that the use and abuse of opioids are commonly tied to the attempt to navigate challenges and to avoid various forms of physical and psychological pain.

The scientific community is gradually coming to understand why people feel pain differently and why experiences of pain change throughout life.[109] This story includes an account of the contributions from genetic differences and an account of the epigenetic factors that modulate the production of neuropeptides and neurotransmitters as well as attitudes toward pain and variations in pain management strategies. But there is a lot we still don't know. For example, the same peptides that are tied to pain management (endorphins, etc.) are altered by both suicide attempts and successful suicide. This suggests that there can be incredibly rapid and pervasive epigenetic effects on the expression of these peptides. We also know that other peptides are implicated in the management of pain, including CRH. This suggests that stress and other ecological challenges are likely to have a physiological impact on pain experience and pain management. Finally, we know that the adaptive function of pain is less prominent when pain is great or when it becomes chronic. As the literary scholar Elaine Scarry argues, pain remains somewhat "inexplicable," though it is a palpable, pervasive, and inescapable feature of human experience.[110]

3

REGULATION

Emotion and Angst

Vincent Dole's early work focused on the fatty acids that underly glucose metabolism.[1] He also held that "drug hunger" was the driving force behind addiction. Specifically, he proposed that a hunger for narcotics was a specific appetite that could not be "satisfied by tranquilizers, sedatives and other psychotropic drugs that also affect the turnover of neurotransmitters."[2] The core of this story focused on the claim that opiate addiction altered the metabolic activity of neurons.[3] The metabolic story might seem like a reflection of a bygone era. But the assumptions that guided this approach persist in neural models of addiction, which often downplay the role of environmental and social factors.

As we have argued in previous chapters, addiction is a form of cephalic dysregulation that is shaped by feedback from the material and social environment, including features of the triggering environments where drugs are used and the social interactions that are associated with drug use. That said, opioid addiction is accompanied by metabolic changes, and it can evoke changes in lifestyle that produce metabolic effects. These changes are not *the cause* of addiction, nor are the ways that poverty and racism impact the brain through physiological responses to stress.[4]

Stress comes in many forms, ranging from limited access to food, housing, education, and employment opportunities to ongoing exposure to violence and other forms of social trauma. Focusing on these phenomena helps to bring the concept of "allostasis" to the fore, offering an alternative understanding of how addiction shapes the brain, body, and behavior.

Homeostatic systems attempt to return to prior forms of stability when they are disrupted by correcting the errors that have occurred. By contrast, allostatic systems adapt—or fail to adapt—to internal and external challenges. In doing so, they use diverse forms of anticipatory regulation—ranging from the secretion of hormones to attempts to participate in or avoid specific social interactions.[5] And while allostatic adaptation is a typically brain-centered process, it often triggers effects that cascade through the brain and body.[6] Collectively, these effects shape and are shaped by the somatic consequences of social and psychological stress—including the consequence of pervasive social inequity.[7] Thus, appeals to "allostatic load" are often used to denote the progressive wear and tear on the body, which results from chronic stress-induced arousal.[8] Likewise, appeals to "allostatic states" are often reserved to characterize physiological responses to stress. But allostasis is much more than this. It is a core principle of neural and bodily design, and it is a process that employs brain-centered, anticipatory processes to facilitate the preservation of viability in changing situations.[9]

Understanding allostatic regulation is essential for capturing the embodied and socially situated nature of cognition.[10] The core idea is that animals must manage bodily and psychological needs while responding to challenges and opportunities that arise in their ecological and social niche. When a physiological state deviates from a predicted need, myriad bodily processes are initiated to preserve viability. This might include secreting molecules like corticotrophin-releasing hormone and cytokines to clot the blood and activate the immune system. It might also trigger the release of hormones such as cortisol, epinephrine, and endogenous opioids to prepare the embodied mind to manage a challenge. But when the levels of these signaling molecules change, their receptors also reset their sensitivity to prepare for future and ongoing challenges. When challenges become more likely, these changes can stabilize in ways that shape future actions and future attempts at physiological regulation. A great deal of research has explored the way that this happens in the context of pervasive social stress, but it also happens in the context of opioid addiction, and this is why addiction is best understood as a form of cephalic dysregulation that depends on numerous forms of anticipation.

The typical pathology of opioid addiction unfolds in three phases: binging and intoxication, withdrawal and negative affect, and preoccupation and anticipation. Each phase is supported by a self-enhancing feedback loop operating at multiple levels. However, the body has no durable set points for opioids. Each apparent moment of equilibrium is a faint echo of homeostasis, which motivates a renewed search and renewed patterns of anticipation.[11] Over time, it takes more opioids to manage pain and more opioids to satisfy increasing desires and compulsions.[12] Furthermore, repeatedly ingesting opioids taxes neural reward systems and contributes to the breakdown of diverse bodily and neural systems.[13] The result is an increase in negative emotional states during periods of abstinence, the endless activation of diverse stress systems, and short-term and decreasing rewards—and all of these factors have a massive impact on the ability to cope with various challenges and opportunities. The drive to avoid pain thus becomes a central aspect of emotional life. Over time, the discomfort associated with physical and emotional pains is exacerbated. The significance of these points is easy to miss. But the same processes that sustain the pursuit of viability also sustain the pathology of addiction. Addiction requires finding ways to access opioids, securing funds for doing so—perhaps by stealing or robbing—avoiding the police, and making adjustments in hopes of finding some semblance of stability. Each of these actions reflects a change to experienced or anticipated needs, and while these actions can turn deadly, they are grounded in attempts to avoid discomfort and to navigate perceived threats to viability.

In more biological terms, we might say that opioid addiction becomes pathological when it bends typically adaptive processes toward maladaptive ends. Some of these processes are involved in managing pain, and others are involved in managing access to necessary resources. This is why diverse neurotransmitters participate in the pursuit of opioids, including dopamine, GABA, glutamate, NMDA, and norepinephrine.[14] These neurotransmitters are tied to the organization of action, the initiation of appetitive behavior, and shifts in attention. Addiction bends the activity of all of the systems that rely on these neurotransmitters toward opioid pursuit.[15] This can promote the devolution of function, but it does so in ways that are tied to seeking a semblance of stability. Unfortunately, this search is often doomed to failure, as the body comes to crave ever higher doses of opioids.[16] Critically, the malleability of these regulatory systems means that addiction is never far away. Fairly minimal changes to physiological systems can rapidly bend thought and behavior toward specific stimuli, shifting the possibilities a person will perceive and transforming the opportunities they will explore. Many of us know someone who is addicted to something, and many of us know what it feels like when there is "never enough."[17] Addiction, however, is not a hijacking of the brain—it is a way that typically adaptive systems are pulled into nonadaptive states through repeated experiences and repeated recalibration of these typically adaptive systems.

ALLOSTATIC REGULATION AND EMOTION

Diverse systems impact opioid ingestion, as well as the experiences of compulsion and withdrawal that are commonly associated with addiction.[18] In many contexts, opioid addiction and withdrawal activate many of the same neural systems that are involved in emotional experience.[19] This goes beyond the claim that addiction is commonly accompanied by melancholic experiences, which increase vulnerability to addiction. Addiction to opioids attenuates hedonic responses, making it difficult to regulate positive and negative emotions. Because opioids modulate the experience of discomfort and the ability to track and respond to threats—including the threat of opioid abuse—these experiences are often flattened in the context of opioid addiction. This can lead to depression and what George Koob calls the dark side of addiction.[20] Just as importantly, anxiety also plays a critical role in the addictive cycle.[21] Many people become anxious before they experience withdrawal, and this motivates them to search for drugs when other resources should be a higher priority.

Emotions are sometimes assumed to be irrational processes that distort cognition and lead us to act against our interests. This perspective sits comfortably with the claim that addiction primarily depends on neural processes that lie beyond the realm of rationality. But in recent years, it has become increasingly clear that this understanding of emotion reflects sexist stereotypes and a tacit denial of our embodied animal nature. Many aspects of cognition are impossible without the contribution of emotion.[22] Moreover, neural damage that impacts emotion regulation also affects decision making, and vice versa.[23] More generally, it seems that emotional capacities are deeply entangled with cognition. Understanding why this is the case offers insights into how affective and cognitive tendencies both become bent toward opioids in the context of addiction.

Darwin famously highlighted the adaptive nature of emotions and the many roles that emotions play in organizing behavior.[24] He argued that emotions are not inferior to deliberative thought, and he proposed that emotions shape the ability to navigate ecological and social contexts in humans and other animals. Contemporary perspectives from neuroscience suggest something similar. From this perspective, it is clear that emotions are typically adaptive capacities, which are woven into the fabric of the mind and brain to aid the pursuit of viability.[25] For example, many of our perceptions of the world and our place within it are shaped by our capacities to respond to the threats, dangers, and opportunities that we are likely to encounter.[26] This is perhaps clearest where perceptions of danger evoke fear and where unpleasant states nudge us to avoid dangerous situations or adopt coping strategies used to navigate threats we cannot avoid.[27] But the ability to recognize emotionally salient phenomena is essential to the pursuit of viability more broadly—if we needed to analyze each potential threat and deliberate about its significance, we could not respond rationally in time.

Experience is commonly structured by "microaffects," which are constantly adjusted against one another as different challenges and opportunities come into view. The physiological processes that produce emotions touch every aspect of cognition through processes that integrate attentional, affective, and visceral information into experience over multiple processing stages.[28] For example, peripheral receptors respond rapidly to the emotional information in bitter tastes, decay-related smells, affiliative touch, and pain; subcortical processes trigger cascades of steroids, peptides, and neurotransmitters as they integrate processing across cortical and subcortical networks; and these effects, as well as many others, continually shape—and are shaped by—diverse forms of perception, judgment, and decision making.

The allostatic processes that make it possible to navigate various challenges and opportunities, structuring these patterns of experience, can be pulled into troubling spaces by individual or social histories. Indeed, as we noted above, the concept of "allostasis" was developed to make sense of the way that physiology adapts to pervasive patterns of stress and trauma, yielding persisting effects on cognitive, affective, and embodied processes. But as we argue in the next section, the impact of changes to the regulatory systems that shape emotional experience is incredibly complex. They often reflect the combined contributions of numerous facts about a person's physiology, the things that they have previously experienced, and their habitual strategies for coping with challenges and engaging with others. This is not to deny that emotional processes can pull people out of habituated responses and tell them that they need to deal with salient challenges or opportunities. After all, moving through familiar and unfamiliar contexts changes the character of emotional experience while simultaneously shaping diverse patterns of thought. Just as importantly, though, people can be thrown into states of fear, anger, social withdrawal, or social need in ways that depend more heavily on their history than on what is currently happening. This is part of what leads to emotions that impede rational decision making or lead us to act irrationally. Critically, this occurs in typical as well as pathological contexts, and it is part of our adaptive capacity to cope with familiar and unfamiliar environments.

Perceptions of emotionally significant stimuli are rapid; they are tied to habituated strategies for navigating challenges and opportunities, and they evoke complex cascades of physiological responses that prepare us to cope with the challenges and opportunities that we perceive. There are numerous reasons why perceptions of danger might be mistaken, but when they are, they can motivate us to act in ways that we later regret. This becomes increasingly likely in complicated situations where it is necessary to deliberate about the long-term effects of an action and where emotions are shaped by habituated biases that are not relevant in a current context. Fortunately, we can typically merge impressionistic responses with slower and more detailed analyses to promote survival and flourishing. Opioids, however, can degrade some of these capacities precisely because of the ways that the degrade responds to pain.[29] Where addictive drives dominate experience, this will often transform emotional and deliberative responses to a diverse range of challenges and opportunities.

DIVERSE STEROIDS, PEPTIDES, AND NEUROTRANSMITTERS SHAPE EMOTIONAL EXPERIENCE

Several neuropeptides, neurotransmitters, and steroids are implicated in addictive experience. Many kinds of search behavior are regulated by neurotransmitters like dopamine and neuropeptides like corticotrophin-releasing hormone (CRH). This is true in typical contexts and in the context of opioid addiction. Addiction affects dopamine regulation in ways that modulate incentive salience, appetitive search, and the organization of action. It also affects the regulation of γ-Aminobutyric acid (GABA), which is tied more directly to experiences of fear and inhibition. It has an impact on dynorphin, an endogenous opioid that is linked to feelings of dysphoria that often surround opioid abuse, and it has an impact on CRH, which is tied to experiences of depression.[30]

Figure 3.1 depicts some of the chemical signaling systems that are implicated in addictive experience and the pursuit and

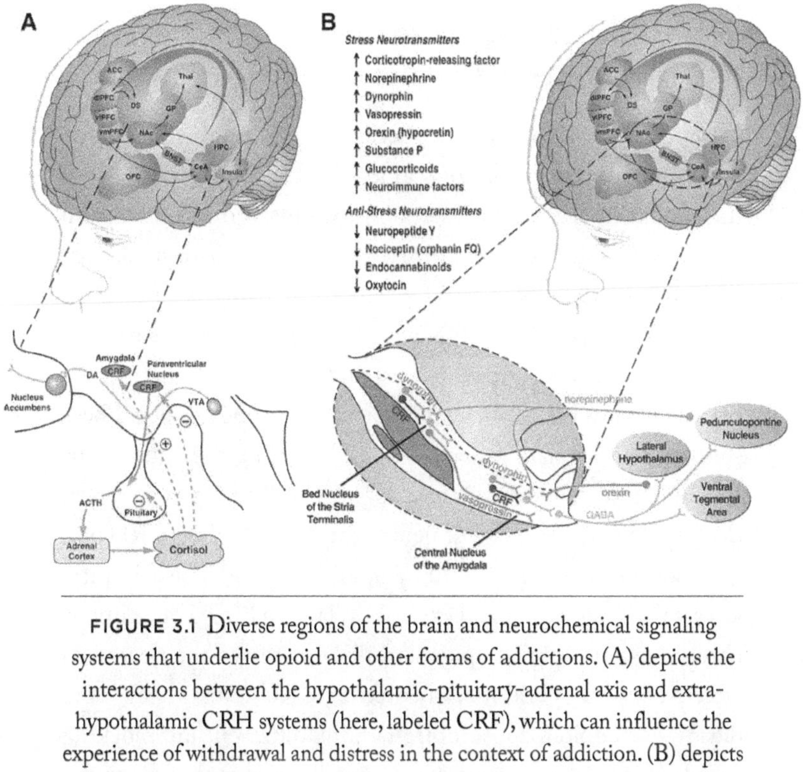

FIGURE 3.1 Diverse regions of the brain and neurochemical signaling systems that underlie opioid and other forms of addictions. (A) depicts the interactions between the hypothalamic-pituitary-adrenal axis and extra-hypothalamic CRH systems (here, labeled CRF), which can influence the experience of withdrawal and distress in the context of addiction. (B) depicts the kinds of allostatic changes that occur in the extended amygdala, which are associated with withdrawal and distress in the context of addiction. The relevant neurotransmitters and neuromodulators are listed to the left of this image. Neurotransmitters systems that convey negative emotional states are indicated by upward arrows; neurotransmitter systems that might buffer negative emotional states are indicated by downward arrows. Reprinted with permission from Elsevier from Koob & Schulkin 2019, 252, figure 7.

consumption of opioids. We cannot tell a complete story about these systems, but we hope to clarify the ways that steroids, peptides, and neurotransmitters shape addictive experience and addictive behavior and provide an overview of the roles that some of these chemicals play in typical and addictive contexts.

Corticotrophin-Releasing Hormone and Glucocorticoids

Neuropeptides like CRH underlie the management of distress. But like other neuropeptides, CRH is quite labile. It is implicated in the appetitive search for salts and sweets as well as drugs of abuse. It is also implicated in the management of distress and attempts to avoid situations that will foster distress. Diverse steroids regulate this neuropeptide in response to various challenges and opportunities, and they do so whether the resource that is being pursued is useful to the body or not. These steroids also regulate other peptides, including the endorphins that underlie the management of physical and psychological distress. The whole range of neuropeptides, including CRH and the endorphins, are impacted by opioids. Opioid abuse, in turn, affects their regulation and expression later in life.[31] For example, chronic use of opioids epigenetically modulates the decreased effectiveness of opioid use.[32] Moreover, CRH levels are elevated during both opioid consumption and withdrawal, and blocking CRH IR receptors can reduce the intake of opioids.[33] But such effects are never simple. Numerous chemical signaling systems shape experience and behavior, and they typically act collectively. Moreover, each of these systems has multiple functions and is entangled with many other systems. Adverse environments impact the expression and turnover of a diverse range of neurotransmitters, including dopamine, serotonin, GABA, and glutamate, as well as neuropeptides such as CRH and oxytocin.

It has sometimes been suggested that CRH is a fear or stress molecule. The release of CRH is often associated with fear, uncertainty, and heightened tendencies to notice unfamiliar and potentially dangerous events. Moreover, animal studies have revealed that administering CRH centrally induces fear behavior, while administering CRH antagonists reduces fear-related behavior.[34] Heightened CRH expression is also evoked in anticipation of adverse and dangerous situations,[35] such as seeing predators or perceiving physiological threats.[36] Opioid use and withdrawal also evoke the release of CRH alongside a range of other neuropeptides and neurotransmitters.[37] But this can't be the whole story. CRH is an ancient peptide that plays numerous roles in development and also shapes diverse forms of adaptive behavior across many contexts.[38]

CRH has long been known to increase salt ingestion, though this could be tied to a sense of adversity that accompanies sodium hunger.[39] However, CRH activation can also increase incentive salience without increasing distress.[40] Furthermore, infusions of CRH into the nucleus accumbens can motivate appetitive behavior while increasing the salience of diverse incentives—ranging from sucrose to amphetamines.[41] This can cement positive or negative reactions, depending upon context, circumstances, and need.[42] Finally, enhanced expression of CRH in the nucleus accumbens has been observed in the context of amphetamine use.[43] In each case, CRH plays a role in actively coping with challenges—sometimes by heightening awareness of threats, sometimes by motivating search and enhancing the willingness to approach rewards, and sometimes by motivating withdrawal from a potential threat. Collectively, these effects organize the cycle of opioid use, abuse, dependence, relapse, and recovery.[44]

The coupling of appetitive effects, incentive salience, and fear responses might seem surprising, but this is possible because CRH affects behavior differently when it is expressed in different brain regions. Studies using optogenetic (light-induced) techniques to modulate CRH activation in mice have confirmed that stimulating CRH genes in the nucleus accumbens and central nucleus of the amygdala promotes increased bar pressing in anticipation of rewards, including sweet tastes, cocaine, and heroin. Stimulating CRH genes in the bed nucleus of the stria terminalis, however, promotes withdrawal and aversion.[45] Simplifying somewhat, this suggests that CRH-regulated activity in regions of the amygdala and nucleus accumbens are tied to the motivational allure of rewards, while their activity in the bed nucleus of the stria terminalis is tied to behavioral withdrawal.

There are at least two partially distinct neural systems associated with CRH regulation, both of which are affected by opioid addiction. The first system is anchored to the hypothalamic-pituitary axis (HPA). The HPA-axis plays a critical role in modulating fight-or-flight responses and managing stress in the context of PTSD, but it is activated in a far more diverse range of conditions where adjustments must be made to the internal milieu in response to challenges. The second system centers on the central nucleus of the amygdala and the bed nucleus of the stria terminalis. This system is more directly tied to regulating vigilance, attention, and awareness, and it plays significant roles in coordinating actions throughout the brain and body.[46] Regions of the frontal cortex employ CRH to regulate the hypothalamic-pituitary adrenal axis (HPA-axis) as well as regions of the amygdala during adaptive and pathological searching, but in the context of addiction, this form of regulation can be bent toward the ongoing pursuit of an addictive substance.[47] These two systems tend to operate independently. CRH levels are elevated in the amygdala and decreased in the hypothalamus during the craving for and pursuit of opioids.[48] But these differences disappear during withdrawal, yielding elevated CRH in both systems.[49]

These factors seem to be implicated in stress relapse, which is a common occurrence in opioid addiction. Stress relapse occurs when someone is pushed beyond their capacity to cope with internal and external challenges. Processes anchored to the bed nucleus of the stria terminalis are implicated in experiences of anxiety.[50] Stress-driven activation of CRH, in this context, seems to be involved in the tendency to relapse under stress.[51] In part, this is because CRH activation triggers an aversive response to withdrawal, which grows stronger with addiction. But critically, prolonged elevation of CRH levels degrades neural tissues and supports some forms of psychic pain and depression that are associated with opioid addiction.[52] We return to this point below.

Glucocorticoids and Other Steroids

These are not localized or isolated effects. They arise through interactions with peptides and steroids and through interactions with features of the ecological and social environment where a person must cope with specific challenges (figure 3.2). Activity in the bed nucleus of the stria terminalis, which is partially regulated by CRH, is commonly tied to the experience of negative affect in the context of substance abuse.[53] This activity is also influenced by glucocorticoids, which shape CRH gene expression, potentially affecting experiences of anxiety and withdrawal, and supporting the negative feedback loops that are commonly associated with opioid abuse.[54]

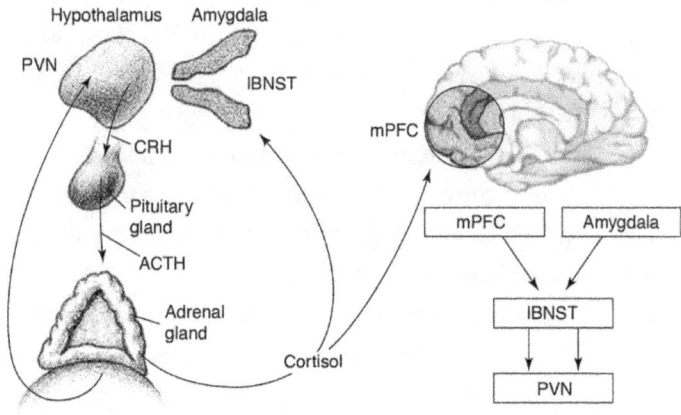

FIGURE 3.2 Schematic image representing some brain regions where CRH is regulated by glucocorticoids. When CRH is released from the paraventricular nucleus of the hypothalamus (PVN), this stimulates the release of adrenocorticotropic hormone (ACTH) from the pituitary gland, which, in turn, stimulates release of cortisol from the adrenal glands. Cortisol can also act on the medial prefrontal cortex (mPFC) and amygdala, providing feedback to the PVN via the lateral bed nucleus of the stria terminalis (lBNST). Reprinted with permission from Elsevier from Schulkin et al. 2005, 633.

Cortisol is a steroid that is associated with the adrenal gland, which commonly affects things like appetite and consumption.[55] In many contexts, the secretion of this hormone is implicated in energy mobilization, but alongside increases of CRH in the amygdala, it also sustains fear in the face of danger.[56] This is adaptive where fear should persist. In typical contexts, the roles of fear and energy mobilization must diverge to regulate different responses to rapidly evolving threats and continued dangers.[57] Cortisol drives the activation of CRH gene expression in the amygdala and bed nucleus of the stria terminalis.[58] It also decreases HPA-axis activation. But the pursuit of addictive substances activates adrenal steroid receptors in the amygdala, nucleus accumbens, hypothalamus, and bed nucleus of the stria terminalis.[59] Over time, opioid search and consumption shapes the activity of this regulatory system in ways that lead to elevated cortisol activity in both the HPA-axis and the system anchored to the amygdala and bed nucleus of the stria terminalis—yielding a persisting experience of angst and anxiety.

Withdrawal experiences are also associated with higher levels of circulating cortisol. However, the prevalence of opioid-related behaviors—including self-administration of opioids—can be decreased in several species by blocking glucocorticoid transcription factors in the amygdala and thereby decreasing CRH activation.[60] In typical contexts, glucocorticoids increase the release of CRH and enhance CRH gene expression throughout the brain, including in the two systems we discussed in the last section.[61] Often, they increase fear- and stress-related experiences along with feelings of compulsion that commonly arise during the pursuit and consumption of opioids. But glucocorticoids are not stress or fear molecules. They are implicated in regulating most neuropeptides and neurotransmitters: they interact with dopamine to mobilize action, they interact with norepinephrine to shape attention, and they interact with serotonin to modulate emotional tone.[62] In adaptive contexts, glucocorticoids also play a crucial role in restraining the activation of the HPA-axis after a challenge has been navigated. This occurs when cortisol levels rise in the blood, and a process known as negative restraint is employed to inhibit further secretion after a challenge has been managed. But importantly, glucocorticoid levels are chronically elevated during opioid search and consumption.

These steroids play significant roles in appetitive search and consumption. Recent data suggest that opioid use and abuse are at least partly supported by changes to CRH gene expression in the brain, which are regulated by glucocorticoids.[63] For example, there are forms of glucocorticoid receptor-dependent plasticity in the central amygdala that mediate opioid addiction-like behavior. This occurs when corticosteroids bind to glucocorticoid receptors (GR) and preferentially recruit the coregulator SRC-1e, activating the expression of corticotrophin-releasing hormone. The resulting increase in the expression and release of CRH during opioid withdrawal then seems to facilitate the pursuit and consumption of opioids. Moreover, blocking glucocorticoid receptors and increasing the coregulator SRC-1a appears to be a way to reverse these allostatic changes in these brain regions.[64]

By contrast, blocking or inhibiting the conversion of corticosteroids in the brain appears to inhibit the consumption of opioids.[65] Here, too, adrenal steroids appear to impact substance use and abuse though the release of CRH in regions like the amygdala and bed nucleus of the stria terminalis. In animal studies, both corticosterone and aldosterone facilitate CRH activation in these structures.[66] Both steroids seem to impact anxious and depressive behavior and the consumption of psychotropic drugs at least, in part, through their impact on CRH.[67] The key thing to notice, however, is that excessive use of opioids can compromise the adaptive role of glucocorticoids in many different ways.[68]

In the short run, elevated glucocorticoid levels might allow someone to navigate the challenges that pervade their life, but the changes to the levels of CRH and other peptides can also sustain the motivational allure of opioids. Over time, persistently heightened levels of glucocorticoids—along with their downstream effects—deplete internal resources and exert a significant toll on the body and brain.[69] Glucocorticoids mediate glucose regulation, inflammation, and metabolism in diverse tissues, and chronic elevation tends to degrade these tissues.[70] During periods of high opioid consumption, they degrade the tissues even faster, as searching for and consuming opioids impacts the expression of diverse stress-related hormones.[71] This effect is then exacerbated by the neglect of normal bodily needs that is commonly exhibited in the context of addiction. So, addiction can make a person age faster by exerting a significant toll on their body.

Each of these phenomena is complex and shaped by a range of interactions between neural and chemical systems. The key thing to notice is that the activation of CRH by steroids primes the search for opioids across numerous different contexts. Cortisol and CRH interactions are elevated during the incentive phase, search phase, consumption phase, and exhaustion phase of opium use. This has longer-term genomic effects and shorter-term effects on membrane-related change.[72] However, all of these changes occur in specific contexts, including the contexts where compulsions and drives have previously materialized. The mere presence of elevated levels of a molecule does not result in behavior, but interactions between steroids (for example, cortisol) and peptides (for example, CRH) contribute to many forms of behavior in specific contexts, including deadly forms of addictive behavior.[73]

These forms of behavior are also shaped by diverse developmental events that impact the expression of CRH and other peptides in ways that sustain increased vulnerability to substance abuse in suitable contexts. For example, where reduced parental support results in diminished meaningful and nurturing social contact, base levels of circulating CRH increase, yielding heightened vulnerability to substance abuse. There are many similar effects that are sensitive to increased glucocorticoid levels in response to ecological and social challenges, which regulate diverse peptides and neurotransmitters in the service of pursuing bodily viability. At least in part, this is because glucocorticoids regulate gene expression of CRH in several brain regions,[74] and the enhancement of CRH can support the pursuit and consumption of addictive substances in anticipation of using them.[75] Finally, many of the processes that are tied to regulating CRH are also entangled with endogenous opioid regulation and expression and, as we argue in the next section, this helps to clarify the complexity of addictive coping and the dark side of addiction.[76]

Endocannabinoids and Endogenous Opioids

Similar patterns of complexity are observed in the context of endogenous opioids. Endogenous opioids are not only elevated during the pursuit of natural rewards but also during the approach or avoidance of resources that can be abused.[77] But while dynorphin levels rise during withdrawal, other opioid peptide levels fall.[78] Pleasure is thus decreased while dysphoria increases. This yields an increased vulnerability to depression and suicide, in part through the downstream activation of CRH.[79] This is the dark side of addiction, which is palpable in many contexts.[80]

At the same time, direct activation of the nucleus accumbens by cannabinoids and opioids tends to amplify the hedonic value of psychotropic drugs.[81] With cannabinoids, this effect depends on the regulation of endogenous opioids.[82] Endocannabinoid receptors are tied to short-term and rapid changes by membrane-related events, which appear to be protective of neural tissue and ameliorative of adversity in some situations.[83] These changes figure in normal signaling for rewards, and they can become compromised by substance abuse and dependence.[84] Cannabinoids are also tied to CRH regulation in regions such as the amygdala, bed nucleus of the stria terminalis, and regions of the prefrontal cortex that are associated with affective withdrawal and reductions of cognitive capability.[85] This matters because appetitive behavior often depends on processes centered on the nucleus accumbens in typical and pathological contexts. The activity of endorphins in this region seems to play a crucial role in motivation and behavioral expression, as well as changes in affect and emotion. Infusions of opioids into the nucleus accumbens amplify the hedonic impact of natural rewards, including addictive and psychotropic substances.[86] Moreover, both opioid stimulation and infusions of dopamine into this structure amplify cue-triggered wanting and motivation to pursue these substances.[87]

CRH is also expressed in many of the same brain regions as endogenous endorphins, and these molecules appear to be tied to experiences of anxiety, despair, exhaustion, and depression.[88] For example, the coexpression of dynorphin with CRH appears to play a significant role in the angst that pervades substance abuse.[89] Moreover, it seems that endogenous opioids are activated by CRH in the context of regulating fear responses.[90] For example, the expression of dynorphin in the bed nucleus of the stria terminalis and basolateral amygdala modulates experiences of fear and anxiety.[91] Moreover, dynorphin seems to be involved in the experience of depression and the dark side of addiction.[92] More specifically, CRH activation of dynorphin and kappa opioid systems shapes the experience of angst and pain sensitivity, and it also shapes patterns of drug abuse and withdrawal.[93]

Expectations, Habits, and Genetic Predispositions?

Returning to some of the issues that we discussed in chapter 2, it is worth noting that molecules like CRH and dopamine are colocalized in diverse regions of the brain. Their coexpression is important for typical forms of appetitive behavior as well as the descent into addiction. These "reward pathways" have been studied extensively in animals, who become obsessed with sending electrical signals to these neural regions, activating the dopamine neurons that underlie psycho-motor expression in context-dependent ways.[94] Here, too, typical forms of adaptive change appear to facilitate the pathology of addiction.[95] Specifically, the obsession with electrical stimulation appeared to be the result of an increased aversion to *not* stimulating. Put somewhat differently, these studies suggested that the loss of stimulation can become aversive, and this suggests that there are cases where animals might self-stimulate not as a way of pursuing reward but as a way of receiving an experience that was salient and anticipated. Unfortunately, the experiences of angst can become deeply integrated with habits and expectations.

Depression is a global experiential state that organizes the inferences we make, expectations we form, and feelings of being overwhelmed or unable to engage with challenges.[96] Sometimes there is good reason to be depressed—because the situation is so grim.[97] Sometimes, it is possible to reframe expectations, increase self-control, and decrease the demands on the brain and body in therapeutic contexts, thereby changing the global character of experience.[98] But this is often more difficult where the activity of numerous neural and chemical systems has been bent by opioid addiction. Many people who are addicted to opioids are vulnerable to depression. This is not surprising given the deeply entrenched expectations of doom and adversity, the pervasiveness of stress, and the tight coupling of habits and expectations to a world where things have commonly gone poorly. But the activity of numerous steroids, neuropeptides, and neurotransmitters are fundamentally tied to this vulnerability as well.

No genes that are directly implicated in addiction have been identified. Twin studies have yielded modest predictions about the vulnerability to addiction, and some candidate genes have been identified.[99] But there is no smoking gun. There is also a small literature—mostly in animal studies—suggesting sex-based, genetic differences in vulnerability to opioid abuse and the preference for opioids alongside differences in pain sensitivity mediated by k-opioid receptors.[100] There is also a literature suggesting gender-based differences in pain perception.[101] CRH is implicated in both contexts, but context, environment, resources, home life, and more remain critical to our understanding of addiction.[102] Moreover, the search for genome-wide candidate genes demonstrates numerous genes and brain regions that are linked to opioid abuse, but they are always tied to other disorders, such as depression, or to circadian processes that sustain the ~24-hour rhythmicity that is essential to behavioral and physiological regulation.[103]

This is significant. While our focus in this chapter has been on brain-centered regulation of challenges and opportunities, regulation always occurs in an ecological and social context. Issues of environment, forms of racism, privilege, sexism, and the threat and experience of jail all affect addiction.[104] People of color are six times or more likely to do jail time than whites.[105] It is hard enough coming out of poverty, but now large swaths of individuals have a prison record that makes their lives even more difficult. Conditions of bio-behavioral exhaustion pervade the brain/bodily sensibility, linking CRH and allostatic regulation to diverse forms of adversity, current and anticipated.

We have highlighted some of the ways that affect and emotion are shaped by a range of neurotransmitters, steroids, and neuropeptides. We have also suggested that addiction relies on the same neural processes that allow us to live at all, and we have gestured toward some of the ways that the global experience of addiction is produced and sustained by processes that bend perception, evaluation, cognition, and emotion toward the ongoing pursuit of opioids. But in asking why some people succumb to opioid addiction, we are also forced to examine the context where they become addicted and the way that their conscious experience depends on the environment where they live and act. Social contexts always impact the regulation of the internal milieu and responses to the social world.[106] This is because challenges and opportunities, histories of stress and distress, and the ongoing demand of navigating physiological needs are always deeply social—and these are the factors that induce adjustments to the networks of steroids, neuropeptides, and neurotransmitters we have been discussing.

Neglecting these facts and treating opium as a drug or medicine that can be prescribed or not makes it easy to ignore the significance of privilege and vulnerability that shape the regulation of the internal milieu, and this has led to pronounced patterns of unfairness.[107] We need to remember that unrelenting adversity degrades physiological capacities unless opportunities for social comfort, social acceptance, social warmth, and the pursuit of social possibilities help to mitigate these effects. Likewise, the endless forms of anticipatory angst, which are organized, in part, by glucocorticoids and the regulation of CRH gene expression in the amygdala and bed nucleus of the stria terminalis, are shaped by engagements with the world and the search for some semblance of stability.[108] This is not to deny the effects of prolonged CRH activation.[109] It is simply to call attention to the fact that these effects occur in a world where we must live and act.

This is significant, as someone who is addicted can often be recaptured not by the drug itself but by other features of the world around them. This can include environments that are associated with opioids. When they interact with places and people associated with drugs, environments can evoke the anticipation of reward that will render them vulnerable to relapse. Just as importantly, novel experiences of distress, uncertainty, or anxiety can also evoke this vulnerability to relapse. Thus, changing the environment, both internally and externally, is essential for managing addictive forms of anticipatory dysregulation. This is often a matter of finding a way to move on—not just mentally, but physically—as we cannot separate cognition from acting in specific environments. We return to this claim in chapter 6. For now, however, we simply want to note how a focus on brain-centered regulation might shape therapeutic considerations.

To begin with, this approach recommends that we situate opioid addiction in the context of the social and psychological contributions to chronic dysregulation.[110] Many pharmacotherapies center the blocking or enhancing of neural processes that shape the pursuit, consumption, and emotional engagement with opioids. This is no easy task. When an adjustment is made by the brain to the complex networks of processes that sustain experience, blocking one aspect of this process can lead other processes to compensate. This necessitates finding additional drugs that might yield a viable point of stability. However, the brain will often adapt to each of these points of stability, making it unlikely that attempting to intervene only on neural processes will succeed as a long-term solution. A more promising strategy would concentrate on restoring social and psychological health, reducing sources of stress and distress, and minimizing conflicts between brain-centered and body-centered regulation.[111] This might also involve working to enhance sleep and exercise, along with the forms of meaningful social contact that are so vital for recovery.[112] Put bluntly, a brain-centered approach to allostatic regulation should remind us that social interventions are often necessary to transform the ways that a person confronts various challenges and opportunities. But what exactly are the social challenges and opportunities that must be addressed in thinking about the social context of addiction? While we can't offer an exhaustive answer to this question, we will offer some landmarks for navigating this complex domain in the final three chapters.

4

SOCIAL HISTORIES AND
SOCIAL CONSTRAINTS

Thus far, we have focused mainly on the neurobiology of addiction. We have acknowledged that social factors must play a role. However, we have not said much about what kinds of factors these are. As we noted in the introduction, we are a socially shaped species that sculpts our social world in profound ways.[1] Consequently, a plausible understanding of the kinds of challenges and opportunities a person must face requires an account of the social and material aspects of the world that they must navigate. To the extent that we can focus on the complexity of these challenges and opportunities, perhaps we can dislodge the sense that some people are condemned to addiction and that others bring addiction upon themselves because they are immoral or irrational.

Across numerous different contexts, overdoses are becoming a more frequent occurrence. More people are becoming more aware that the warm embrace of oblivion can give way to cold lifelessness and the collapse of respiratory systems.[2] However, many questions remain about how patterns of addiction are embedded within and shaped by cultural contexts. Fortunately, the suggestions that addiction is a form of irrationality or a form of immorality are no longer expressed as if they are obvious. Moreover, it is harder to defend the claim that addiction is a vice or an artifact of "inner-city drug use." It has become much harder to deny that the use and abuse of opioids are just as common in small towns and rural areas as in large cities. But we have a long way to go before we have dislodged the biases and assumptions that can make it seem like someone is the "wrong kind of addict."

Our aim in this chapter is to explore some of the ways that cultural factors shape the desire for opioids and the likelihood of addiction. Historians have often noted, however, that we cannot understand opioid addiction from a single perspective. There are multiple ways for people to become addicted to opioids, and there are multiple social and existential phenomena that shape patterns of addiction. Critically, the social and cultural forces that shape addiction are also incredibly complex. We thus present several snapshots of different cultural contexts where opioids have been used, with different assumptions about what it meant to use them. Specifically, we offer a high-level overview of the ways that ancient religious practices, modern scientific explorations, war, colonialism, corporate interests, and social narratives have organized the social landscape where the current opioid crisis has taken hold. To be clear, however, we have no doubt that some people have become addicted in each of these cultural contexts. Moreover, we acknowledge that many details must be glossed over as we attempt to highlight some of the interacting factors that shape the social landscape where addiction becomes possible. But we hope that the differences among these cultural contexts can reveal some of the many ways that cultural constraints increase or diminish the likelihood of broadscale opioid use—and we hope it becomes clear how these constraints shape the prevalence of opioid addiction.

LEARNING TO USE OPIUM

Across a wide range of ecological contexts, humans have learned to recognize plants with nutritional value and plants with psychotropic properties, including tobacco, coca, betel nuts, and poppies. Like all other animals, we try to survive. When our basic needs are met, we also pursue opportunities to live better.[3] We explore new foods by sprinkling them with something safe and familiar.[4] We also increase the hedonic features of the food by adding spices and other flavorings, and we keep track of things that can be used to manage pain. Perhaps people stumbled upon the power of poppies as a side effect of ingesting them, or maybe they inhaled smoke from burning poppies and noticed the result. Once they discovered that poppies could reduce physical and psychological pain, however, the value of this plant would have increased, and it would have been categorized in terms of its contribution to health and well-being.

The systematic search for painkillers stretches beyond recorded history. But this search probably reflected a pervasive need to relieve pain in contexts like hunting, managing digestive and inflammatory diseases, and coping with the effects of warfare.[5] We cannot say for sure when people realized that opium poppies could be used to address diverse kinds of pain, but each community that has experienced opium appears to have found ways of using it to treat various forms of discomfort.[6] Even where it has been impossible to treat the underlying causes of disease, people have recognized that opium reduces suffering. However, while opioids might help people feel like they are thriving in the face of challenges to viability, they can quickly become the reason that people do not.

The immediate effects of ingesting opium led to its adoption as a component in many products in prehistoric as well as historic times.[7] But the long-term effects have often yielded numerous social complexities. Moreover, the tradeoff between the short-term and long-term effects of opioids has always been a source of social complexity. Written records suggest that poppy seeds have been used as food for at least four thousand years, and there is ample evidence that poppies were used for healing, magical practices, and recreational calming throughout the ancient world.[8] For this to be possible, robust agricultural practices were necessary to allow for the cultivation of poppies in a wider range of contexts. Moreover, improvements in storage and harvesting were necessary to sustain the distribution of the poppy. Once these factors took hold, the poppy spread throughout the world.

The initial use of poppies is likely to have started in the Eastern Mediterranean, but it rapidly spread throughout Asia and Europe. The ancient Sumerians might have called the poppy the "joy plant" (though this claim is contested). However, there is ample evidence that poppies played significant roles in Egyptian medical and magical practices.[9] The earliest image of opium poppies in an Egyptian tomb is approximately 3,200 years old (figure 4.1). Older medical tests describe uses of opium to manage diverse ailments, and traces of opium have been discovered in jars that were imported from Cyprus.[10] Unfortunately, it is

FIGURE 4.1 A depiction of red poppies from the tomb of Sennedjem (ca. 1295–1213 BC). Adapted from Sennedjem and Iineferti in the Fields of Iaru. Metropolitan Museum of Art. Public domain image.

difficult to get a handle on the earliest history of this plant's cultivation. What is clear, however, is that it rapidly spread to different communities as similar accounts of its use began to emerge throughout the ancient world.

In part, this is because poppy seeds were cultivated in Central and East Asia, and the knowledge that was needed to cultivate the opium poppy in a broader range of ecological contexts was disseminated across the expansive silk and spice trade routes. Chinese and Indian doctors thus developed diverse understandings of the plant as a cure-all for thousands of years before doctors in Europe and North America caught on.[11] As a result, the discussion of opioids became more common within traditional medical texts. Even early on, there was a sense of both opium's potential to heal and its potential to give way to addiction and abuse. For example, it was acknowledged that those who fall under its sway would often find the substance worth pursuing with relentless zeal. This brings us to a key point: medicinal practices never exist in a vacuum, and decisions to use a particular approach to pain management always depend upon what alternatives are available and what priorities are privileged in particular contexts. Even in ancient contexts, we find discussions of alternatives to opium in the same texts that discuss their power, and we find that opium use is not simply presented as a matter of managing pain—it is also presented as a vector for cultivating subjective well-being or distinctive forms of spiritual experience.

Consider the ways that opioids were discussed by Greek philosophers, storytellers, and ancient doctors like Hippocrates and Galen. They called attention to the utility of opioids in medical treatment, and they acknowledged that using this substance could easily slip into abuse. Much like today, people knew that the use of the poppy could develop a life of its own. Intriguingly, this seems to have resulted in an idealization of the poppy in many contexts across the ancient world. Egyptian and Greek statues were erected to honor the poppy. There was a poppy goddess in Gazi, Crete. Aristophanes alludes to the ancient drug in *The Birds*, treating the poppy as an aid to the human condition.

For the Greeks, the poppy belonged to three gods: Thanos, the god of death; Hypnos, the god of sleep; and Morpheus, the god of dreams.[12] Morpheus was a god who pulled people out of their misery and led them toward the underworld. When opium is ingested, it can evoke a sense of musing with the gods, and it can allow the mind to escape into a space that feels dreamlike or like a kind of mystic harmony. Perhaps this is why the muses of ancient Greece were thought to be aided by medicinal herbs, including opium. In any case, the modern drug morphine is named after Morpheus in an attempt to preserve some of this resonance. Aristotle or Seneca might have agreed with these ways of thinking about opium, though they would have added "in moderation." But this is precisely the problem. It is hard to be moderate in response to psychic or physical pain, and there is nothing to block the possible slip from recreational use into decadence and decay.

SCIENTIFIC EXPLORATIONS

The development of mathematics allowed people to construct impressive material structures. Likewise, the prevalence and accessibility of written language facilitated the construction and codification of explicit medical principles and memorable medical narratives. As we noted in the introduction, the ability to record knowledge played a significant role in the way that people understood substances like the opium poppy. Specifically, narrative presentations allowed people to move from the discovery of the poppy to a diverse range of strategies for communicating about its properties and eventually to the codification of knowledge about this substance to be shared with people in distant places and times. The stories people have told and the ways they have understood opioids have had a deep impact on the culture of use and abuse.

There is a long history of mixing opium with wine for medicinal and recreational purposes, as well as magical and religious practices. This appears to have been the core of the Egyptian practices centered on the use of the poppy, and it was central to the ways that the poppy was used in ancient Greece. In these contexts, the soporific properties of the poppy were well-known. But in the sixteenth and seventeenth centuries, the use of the poppy once again became familiar enough to play a prominent role in popular culture. Thus, in *Othello*, we find Shakespeare mentioning the powerful effects of opium after Lago has upset Othello so much that sleep seems impossible:

Not poppy nor mandragora
Nor all the drowsy syrups of the world
Shall ever medicine thee to that sweet sleep
Which thou owedst yesterday.[13]

Over the next several centuries, the discussion of opium began to diversify as practices of scientific exploration and medical experimentation came to the fore. For example, in early modern Europe and in what is now known as North America, people cataloged plants, searched for novel varieties, and explored their potential uses. The result was an emerging approach to natural history, which integrated narrative imagery with medical insights in ways that could be more easily consumed and remembered.[14]

For example, Erasmus Darwin—a poet, physician, and pious naturalist—presented several critical insights about medicinal herbs in his poem, *The Botanic Garden*.[15] As a poet, he articulated a deep love of plants, but as a physician and a naturalist, he also articulated numerous medical and nutritive insights—including his understanding of the opium poppy. Here, poetry and botany converge as Darwin presents the good and bad features of the poppy. Importantly, this allowed him to explore the human condition both in the context of dreamlike states induced by opium and in the context of medicinal strategies for ameliorating pain.

Throughout this period, robust and systematic taxonomies were imposed upon the dizzying array of plants that were being "discovered" by Europeans. Of course, natural history has always been deeply embedded in cultural history, but during this period, a sense of discovery, ownership, and control began to take hold, driven in part by colonization and the emerging scientific revolution. By the seventeenth century, scientific cultures were relatively stable and well-supported, allowing scientists to gain a greater appreciation of the powers of various plants. The English physician Thomas Sydenham held that opium was a God-given remedy for suffering, which should be used to manage all forms of discomfort and pain. To pursue this goal, he developed a recipe for laudanum, a mixture of opium and alcohol that could be used medically.[16] From this point forward, the use of opium in science and medicine became increasingly common.

The opium poppy served as an exciting context for medical "discovery": in small quantities, it was thought to play a medicinal role, but in larger quantities, it was understood to yield "intoxication, languor, stupor and death."[17] One early attempt to confront the problem of addiction is found in the work of Benjamin Rush, who is often designated as a founding father of the United States. Rush was a well-known physician. He

was also a reformer who wrote pamphlets advocating temper-
ance and highlighting the fact that alcohol abuse often led to
morally problematic behavior.[18] In his youth, Rush was forced
to navigate engagements with a father who drank excessively, so
he was deeply familiar with many of the dark facets of substance
abuse. But as a doctor, he supported the use of laudanum—
while acknowledging its dangers—for managing yellow fever
and other diseases. We note this because even the most cau-
tious doctors could not deny the significant benefits of opium,
even though they knew that it was possible for people to become
entangled with their effects.

The synthesis of morphine opened up a great deal of space for
further medical experimentation. In 1804, Friedrich Sertürner
chemically isolated morphine, facilitating the mass production
of opioids. Medical research on opioids has a complex history,
which we will not detail here.[19] For our purposes, the key thing to
note is that Francois Magendie initiated the experimental exploration of morphine in the early nineteenth century in an attempt
to articulate its principles of action. He also produced the first
pharmacological textbook to be based on scientific observations
and investigations. One of Magendie's students, Claude Bernard,
used injections of opium as an anesthetic in the experimental
contexts where he was exploring diverse physiological phenomena. His research also famously led him to articulate claims
about the centrality of regulating the internal milieu to the pursuit of viability.[20] Bernard's approach continues to serve as the
conceptual and empirical foundation for scientific understandings of homeostasis and the allostatic approaches that we have
explored throughout this book. Critically, however, they are the
conceptual foundation of claims about the "wisdom of the body,"
they guided the initial exploration of the chemical signaling
systems we have discussed, and they provided a framework for

how they might integrate diverse systems throughout the body.[21] Subsequent research in this tradition made it clear that bodily needs can be determined through physiological and behavioral cues, revealing clear pathways to the management of deficits and to the restoration of homeostasis.[22] While some people would eventually come to disagree, these ideas were quickly integrated into the medical understanding of opioid use and abuse.

This is one of the main things that led doctors in the United States and Western Europe to supply opium, morphine, and heroin to people who had access to healthcare in an attempt to manage various forms of pain.[23] They recognized the costs of doing so, but they also thought that it would be possible to manage "the opium appetite" by satisfying the continually increasing need for opioids.[24] From our current perspective, it is clear that this was a bad idea, but morphine maintenance and heroin maintenance were both initially viewed as plausible therapeutic interventions. Strikingly methadone, naltrexone, and buprenorphine maintenance programs have emerged more recently as part of a similar attempt to treat the scourge of their predecessors.[25]

The maintenance of opium dependence has often paralleled the use of insulin therapy for type-1 diabetes. We give insulin to type-1 diabetics whose pancreas cannot produce the insulin necessary to manage everyday glucose regulation. In a similar way, it has often been assumed that opioids should be given to people who are unable to regulate opium cravings.[26] But the physiological demand for opioids doesn't fit comfortably within a homeostatic theory. It looks like a need, but the "need" for opiates is different from the need for insulin—it can destroy a person over longer time scales.[27] While it has sometimes been assumed that opioids must be employed to bring the body back into a state of balance, this practice has not led to successful and general strategies for managing opioid addiction, nor has it always been accompanied by the requisite cautionary notes about the potential for abuse.[28]

This is somewhat surprising. But there was a broad culture of "medical mentalism" in the nineteenth century.[29] Many medical professionals were apprehensive about opioids because they were aware of the advantages of opioid medicines, but they also knew about the dangers of becoming addicted to them. But during the nineteenth century, as many as 40 percent of doctors in Germany might have sampled their stocks and become addicted.[30] Some scholars disagree with this estimate, even where the data are relatively clear. But there is a broad consensus that the proportion of doctors who sampled the opioids that they had access to was both high and alarming.[31] Perhaps the explanation for this lies in the vulnerability of everyone—including medical practitioners—to the allure of opioids under the right circumstances. Put bluntly, many physicians and others who "should have known better" often ended up with complex practices and commitments that made it hard to know what to think about this newly medicalized substance, and this led to a wide range of confusing signals that made it difficult for anyone to know exactly how to proceed.

Some medical practitioners presented opium in a positive light, as emboldening warriors, reducing pain, and maximizing resilience.[32] William Stewart Halsted, one of the founders of the Johns Hopkins Medical School, regularly used opioids to dull psychic and physical pain. He never overdosed, and he could always sustain his habit, as he had nearly unfettered access to opioids. Moreover, he remained an awe-inspiring surgeon, but his excessive use of opium and morphine was well-known.[33] William Osler, another founder of the Johns Hopkins Medical School, offered warnings about opium addiction, highlighting the use of hypodermic injections by physicians in *The Principles and Practice of Medicine*.[34] He also told his students that they should cultivate a detailed knowledge of opium, "God's own medicine."[35] Finally, Oliver Wendell Holmes Sr., another nineteenth-century dean of medicine, vehemently opposed opioid treatment and government involvement in regulating opium.[36] However, by this point, corporate interests were becoming entangled with broader patterns of opium use and abuse, and this rendered the conflicting judgments of medical professionals almost meaningless.

The situation would only become more complicated when heroin was synthesized by C. R. Alder Wright in 1874 and again, independently, by Felix Hoffman in 1897. Hoffman worked for Bayer, the German pharmaceutical company. It is likely that the name *heroin* derives from *heroisch*, reflecting an attempt to frame the substance as a "heroic" resource.[37] Heroin caught on quickly, and it became a common drug for managing headaches, pains, colds, and other ailments like tuberculosis and pneumonia.[38] It was prescribed by doctors as a miracle treatment for a wide range of ailments, and it was readily available for home use (figures 4.2a and 4.2b). Perhaps most surprisingly, an opium-based product known as Craig's Heroin Compound was used to manage infant discomfort.[39] As infant mortality rose to catastrophic levels, such practices were questioned.

It is not clear if heroin was a major contributing factor to infant mortality, but the Pharmacy Act of 1868 was enacted in the United Kingdom to address this crisis, bringing opium under state regulation for the first time in Western Europe.[40] Perhaps unsurprisingly, clinics continued to sprout up across the United States. Narcotic use became increasingly common, addiction began to affect a range of people from myriad social contexts, and heroin began to slip beyond the medical context, allowing patterns of addiction to expand beyond medical usage to spaces of recreational use and abuse.

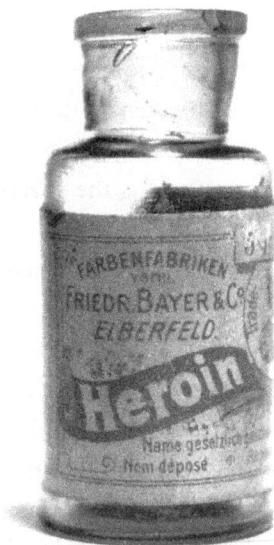

FIGURE 4.2A Bayer heroin in a five-gram bottle. Wikimedia Commons.
Public domain image.

FIGURE 4.2B Craig's heroin compound, being promoted for treating
infants. Reprinted from *Interior Journal*, Stanford Kentucky, April 10, 1906.
Public domain image.

Of course, many doctors have recognized the cycle of maintenance and withdrawal, and this has often led to an understanding of addiction as a disease. Moreover, scientists have often tried to dampen the high and create substances that can manage pain without leading to addiction.[41] To this day, research has continued to center on maximizing the effectiveness of opioids by magnifying some of the properties of the opium molecule and altering parts of its molecular structure.[42] Tragically, this goal has often been beyond the reach of the best existing science, and the magnification of pain management has often increased the addictiveness of opioids. This is perhaps clearest in the case of synthetic opioids, such as fentanyl, where the result has proven lethal and destructive—but this is an instance of a broader historical pattern. More recently, and more familiarly, Oxycontin was developed and marketed as an opioid without the addictive potential.[43] Caution was urged from the beginning, but the worries were overwhelmed by Purdue's marketing—physicians were actively recruited to promote it, and advertising campaigns were directed toward physicians as well as other healthcare workers. Multinational cartels also continue to provide illegal opium, with and without the explicit backing of state power. Across all of these contexts, the violence and wealth that is generated by providing legal and illegal opioids are not so different from the supply of these substances in the nineteenth century. This has always been a big business, and that's part of the reason why it has been so difficult to get traction on every opium crisis.

WAR AS A CONTEXT FOR OPIOID USE

A recurring social theme that arises across numerous historical contexts is that addiction rates often increase in the wake of attempts to manage the pain of war. Opioids have often been used to keep soldiers fighting and to mitigate the physical and psychological trauma that persists in the aftermath of war.[44] When substances like codeine and morphine were synthesized during the first part of the nineteenth century, they quickly became common tools for the relief of pain in the context of war. Opioids were routinely given to wounded soldiers in both the American Civil War and the Franco-Prussian War. Often, pain medications were available even when other necessary supplies were not. Here, too, however, access to opioids and the understanding of their use differed substantially across social contexts.

In a letter to Sir Frederick Pollock, Oliver Wendell Holmes Jr. recalled being wounded during the Civil War. He wrote that he had kept a bottle of laudanum—the liquid mixture of opium—in his pocket and that he had "resolved if the anguish became unbearable to do the needful." Before he used it, however, a doctor removed the bottle from his pocket, and Holmes resolved to live in the morning.[45] For Holmes, the bottle of laudanum was a secret for managing pain and confronting death if necessary. He was a privileged individual caught in a worthy fight. The great Spanish neuroscientist Santiago Ramón y Cajal also wrote of the "inevitable bottle of laudanum" that he carried as a physician in the Spanish army to treat diseases such as tuberculosis and diarrhea.[46] Finally, in the context of the Crimean War, Florence Nightingale commonly used opium to treat people who had been wounded, and many doctors followed suit. The promise of opium led to increasing the use of opioids for many kinds of ailments, even where they didn't solve the underlying problem. As a result, many people ran the risk of becoming addicted to opium without their underlying maladies ever being addressed.

War has always been a breeding ground for opioid addiction, and complex interactions have often emerged between the use and abuse of opioids and warfare. Attempts to ameliorate physical, social, and existential pain often transform broader patterns of opioid use and abuse. In the context of the American Civil War, some people referred to laudanum as "Saint Morphine."[47] The rhetoric quickly changed as broader patterns of use and abuse came into view. Opioids were used to minimize the pain of lost friends, lost limbs, and degraded dreams, but this expanded use of opioids was often the first step along the path to addiction, and addiction became a problem for many Northern soldiers.[48] Opium abuse was also commonly observed among imprisoned soldiers. In these contexts, use of the drug was often seen as a moral failing or an expression of weak character.

Opioid use was not limited to soldiers at this time, however. Across diverse social actors, it was used in ways that reflected the kinds of pain that commonly follow in the wake of war. For example, David Courtwright notes that opioid abuse was widespread in the aristocratic class of the defeated South.[49] Specifically, doctors seem to have overprescribed opioids to Southern women, who were beholden to a "lost cause," allowing them to self-medicate as a way to minimize their experiences of mental distress.

The struggle with opium addiction permeates Southern literature (for example, the elderly Mrs. Dubose in *To Kill a Mockingbird*). It can also be observed in the writings of Mary Chesnut, the well-known Civil War diarist from South Carolina. Like many others, she appears to have "relieved the tedium by taking laudanum."[50] Then, as now, the management of psychic pain meant killing off part of the person. Importantly, though, these uses did not occur in a vacuum. Opioids were considered a good remedy for many illnesses, and they were given to numerous men and women with the best intentions.[51] Perhaps most importantly, in this context, opioids were seen as an aid to flourishing and not evidence of a moral failing.

Our understanding of this situation cannot be so simple. Everyone loses something on the battlefield, and abuse often spikes after a war. War ensures the use of opioids to manage pain and discomfort, and it also ensures access to the drug, a culture that permits its use, and numerous other forms of trauma for which opioids promise an escape.[52] Furthermore, once opioids take hold of a population, their use is rarely constrained to the management of pain and discomfort. Often, there are people who benefit from the patterns of use and abuse that follow in the wake of war and trauma. This is a broad historical pattern, and it is seen across many different contexts.[53] The "morphine habit" became a "blessing" when it reduced suffering and a curse that one had to keep an eye on to "combat its evil effects."[54] As such, some doctors began to warn of the dangers of opioid abuse, while many others contributed to it. Here, too, there were larger issues at play.

COLONIALISM AND INCREASING ACCESS

In the earliest days of scientific exploration, experimentation with opioids led to an explosion in the use and misuse of opioids. This, in turn, required an expansion in the supply of opium poppies, which resulted in massive economic and social pressure on places like Bengal, which were integral to the opium trade. Scientific exploration thus lent support to the entrenchment of colonialism. Further political pressures on the use and misuse of opium came into view when the British discovered ways of exploiting Chinese practices of smoking and ingesting opium. Even as the British government was awakening to the dangers of opium use, they were scaffolding their empire with opium profits. Like the "discovery" and exploitation of tobacco, this "opportunity" would shape the world in profound and awful ways.

A diverse range of striking examples of the integration of greed and imperialism can be found in this historical context. For example, the Chinese government reacted with alarm to the rising levels of addiction, supported by the increase in opium trade under the East India Company, and they attempted to limit opium's availability in light of this danger. This was a rational decision, which anticipates our own government's attempt to prevent addiction from spiraling out of control by limiting availability and punishing people for using opioids.[55] This attempt to regain social control threatened British imperial power, however, and it limited the profits the British East India Company would receive as a result of importing South Asian opium into China.[56] To safeguard these profits, the full military power of the British state was turned against China as a way of preserving their profit margin in the face of a growing opioid epidemic.

The connections between social factors and opium addiction are rarely straightforward. As such, we should tread cautiously in thinking about the link between political or monetary interests and the promotion of opioid use and abuse. The pursuit of global trade—including trade in opium—also led to the creation of great wealth in the United States, sometimes by exploiting people and their most destructive impulses.[57] The rise of pharmaceutical companies in Europe and the United States, for example, provided an excuse to expand the market for opioids to the citizens of the imperial powers. In the mid-nineteenth century, Russell & Company was one of the largest American trading houses in China. One of the directors of this company, Warren Delano Jr., appears to have made a great deal of money from the opium trade, and he may have played a significant role in the Opium Wars. Delano was the grandfather of Franklin Delano Roosevelt, and it seems likely that the accumulation of wealth from the opium trade played a crucial role in the establishment of the Roosevelt family fortune.[58]

There are plenty of other cases where the connections are less clear. Consider John Stuart Mill, one of the most famous philosophers in the United Kingdom in the nineteenth century. Mill was an active member of the East Indian Trade Company, and his career was supported by the opium trade, as opium was one of the most profitable products for the company. Mill's work with the East India Company was entangled with the pursuit of imperial interests and profit from opium, and it demonstrated that catering to the interests of specific people can harm both society and individuals. By contrast, his philosophical perspective was an attempt to reconcile the maximization of happiness for the greatest number of people with freedom of expression and respect for the preferences of others. Figuring out where to stand on the relationship between these perspectives is no easy matter, and this is a common situation in contexts where colonialism shaped the trade of opioids and entrenched social priorities anchored to the maximization of profit.

With these facts in mind, we can begin to see traces of the complex networks of interactions among medical treatment, social location, and addiction that shape the rise of opioid epidemics. These issues are incredibly complex. Many doctors hope to ameliorate or reduce suffering.[59] Many people pursue self-improvement and increased autonomy, and some people unscrupulously chase profits. But the chasing of profits from opioids has a tangled and global history. Sometimes, this benefits individuals, and sometimes, it benefits corporations, but in every case, it is the people who must find a way to live in the wake of these decisions who bear the brunt of the trauma.

Opium-containing medicines were used for centuries in China without yielding broadscale patterns of addiction. Patterns of recreational use had long been present. The context of opium use shifted in the nineteenth century, however. As a result of the two Opium Wars, the trade of opium increased radically, peaking at 15 million pounds in 1879—this was far from the end of the story, as the southwestern provinces in China were also producing an average of 32 million pounds of opium to satisfy the increasing demand.[60] At the same time, various forms of political and cultural pressure within China led large numbers of Chinese immigrants to come to the United States. Many of these immigrants worked as mine laborers, lumberjacks, or laying railway lines. They were often single men, many of them came from social classes with access to fewer resources, and when they arrived, they experienced pervasive forms of exclusion, oppression, and racism.[61] They also brought the recreational use of opium to the United States, and they introduced many people to the opium den (figure 4.3).

FIGURE 4.3 Opium den in a Chinese lodging house in San Francisco, California (circa 1890). Wikimedia Commons. Public domain image.

Initially, opium dens tended to be ignored because opium use was primarily thought to be confined to Chinese immigrants—and racial biases led people to be less concerned with the harms that occurred in primarily Chinese spaces. As growing numbers of white people explored these spaces, societal concern about opium use rapidly increased. By the early twentieth century, laws were enacted against the smoking of opium. At this point, it was too late. As opium use became illegal, the danger surrounding its use increased in ways that stretched beyond the physiological and psychological effects on the smoker. Organized crime encroached on opium dens. Criminals became the primary suppliers of opium, and as opium trafficking came to be recognized as an international problem, academic conferences sprung up around this emerging international scourge.

Of course, there have always been people who have expressed skepticism about attempts to regulate drug use. Supreme Court Justice Oliver Wendell Holmes Jr., for example, proposed that the institution of laws surrounding drug use was likely to create more criminals without addressing the broader issues (*U.S. v. Jin Fuey Moy*). He also expressed doubts about the attempt to regulate the prescription of morphine for maintenance purposes (*Webb v. U.S.*).[62] Just as importantly, legislators saw an opportunity to capture revenue from people who had unshakeable habits, so they decided to tax both the consumption and the production of opioids.[63] As a consequence, opioids were not banned when they first started to cause social problems in the United States. Even when the Harrison Narcotics Act of 1914 made the use of opioids illegal, it was still possible to obtain heroin from doctors. Social elites have always had the best access to medical care for their pain, and this situation was no different. Moreover, the understanding of drug use has not extended equally to everyone who suffers. This supports large-scale patterns of differential concern, as well as disparities in access to medical and pain management resources, which continue to raise deep ethical questions of what we owe one another.[64]

In this context, some people retained access to "medicine," while others were forced to pursue other means of obtaining opioids. Those who were under the care of a physician were treated as legitimate users who deserved compassionate support. But there was a great deal of room for illegal sales, which fostered a wealth of social complexities. Just as Prohibition had failed to eliminate the excessive use of alcohol, the outlawing of morphine sales did not end its use. However, those who pursued nonmedical access to opioids were treated as criminals; those who sought out doctors who were willing to provide them with access to opioids were not. We should all be incensed at the way poor people, who are least able to access healthcare, not only suffer the most but are also the most susceptible to the criminalization of addiction. The management of pain should not be a privilege for just the few; it should be an imperative that applies equally to everyone. But this isn't how things tend to work.

Physicians and other healthcare providers continue to supply a range of opioids, sometimes intentionally, sometimes as a result of poor judgment, along with social and corporate pressure. Often, this supports the conditions that give rise to broader patterns of addiction. In most cases, the public expression of apprehension is unable to limit the spread of opioid use and abuse. To a large extent, this is because Big Pharma tends to look for ways to exploit the initial use of opioids and the possibility of therapeutic interventions.[65] But this situation is complicated by the fact that access to opioids often comes to operate in illegal and legal ways. Even where opioid use begins with a doctor's prescription, addiction sometimes leads people into situations where they must try to obtain more by stealing or turning to other forms of behavior they would not otherwise consider. As the experience of addiction progresses, people often learn to lie, hide their behavior, and attempt to preserve social legitimacy. People of all ages, from all walks of life, can find themselves chasing a high, stealing for it, or deciding to acquire opioids by any means necessary. Across all of these contexts, control is sought. The state prohibits some forms of access to opioids and punishes some people who are caught up in the pursuit of opioids, but the distribution of support and punishment has never been fair or equal.

THE EMERGENCE OF RICH CULTURAL NARRATIVES

There is one final strand of cultural phenomena that is worth considering as we think about the nature of opioid use and opioid addiction. Many famous scientists, writers, and philosophers have discussed their use of and experience with opioids. While it is hard to know how these kinds of texts have impacted the use and abuse of opioids, they do provide insights about broader cultural trends and the kinds of factors that have been at play when people have engaged with them.

Unease surrounding the medical use of opium has always been palpable. This is not surprising since the addictive side of opium cannot be ignored. Many people have been rationally apprehensive about the use of these substances, and this is part of the reason why scientists and pharmaceutical advertisers have often portrayed such substances in seductive ways, for example by using Morpheus (the Greek god of fantasy or dreams) to give it a more positive connotation. Sometimes, they would con textualize the benefits of relief from pain by appealing to the dangers of abuse and the possibility of becoming caught up in the demands of the drug. However, these claims have always received complex forms of support from physicians and scientists, and they have always needed to be contextualized within the broader context of depictions of the experience of using opioids. Here, too, however, patterns of unequal access have often shaped the way that using and abusing opioids is portrayed in the social imagination. This has yielded an uneven understanding of what the experience is like and what its status is.

During the nineteenth century, for example, opium diaries came to prominence in Europe and North America.[66] Perhaps the most famous of these diaries is Thomas de Quincey's *Confessions of an English Opium Eater*, which alludes to the alluring properties of opium alongside the complexities of ongoing use.[67] Many others followed suit, attempting to explain the good, the bad, and the ugly of opium ingestion and obsession. People described how sleep opened into dreamy states, allowing for a retreat from the present. They also noted that this pathway could lead to nightmares, immersive feelings of melancholia, or experiences of distorted possibilities.[68] In general, however, opium use was glorified in memoirs and autobiographies throughout the nineteenth century. These were depictions of romantic oblivion that depicted people with time and money to pursue experiences of oblivion.[69] Importantly, they were rarely just descriptive—they were also confessional. They were depictions of a world of possibilities that gave way to the delusional allure of a warm high, and sometimes, they revealed the dark side of addiction. These aspects of opioid use and abuse would also shape the discussions of the opium dream that saturated various forms of literature and poetry in the nineteenth century—for example, the works of Baudelaire, Coleridge, and Keates. David Courtwright discusses these trends in detail, but the key thing to notice is that melodies of discordant expression and atonal bombast saturate the accounts of use and abuse during this period.[70]

Few other substances can protect a person from physical pain, placate psychological discomfort, and stave off aversive experiences. This makes it easy to understand the allure of opium and makes it hard to draw any substantial conclusions about how public reflections on opioid experiences affected the availability and increasing use of these substances throughout the nineteenth century. One thing that is clear, however, is that public reflections on the harms of opioid addiction often come too late, when broadscale patterns of opioid dependence and addiction have already set in. Strikingly, discussions of the vulnerability to abuse were also written during the nineteenth century alongside the more glowing depictions of the opium high. These had a much more minimal impact on the cultural imagination. In a poem that remained unpublished during his lifetime, George Crabbe provides one of the most realistic warnings of the power of the poppy, while avoiding mythologizing opioids:

> Deep is the Poppy's blushing red
> Ah! take it from our joyous bowers.
> With baleful Dew its flower fed,
> Until, replete with deadly powers,
> Its heavy influence round is shed,
> That ease and cheerfulness o'erpowers.
> No being loves it, all would hate
> Did it not men intoxicate.[71]

Of course, history is also littered with stories about people who were plagued by drug abuse. In the nineteenth century, Charles Sanders Peirce suffered from numerous ailments, including facial pain, or trigeminal neural irritation, and he self-medicated with opium.[72] He was repeatedly warned that his use of this substance might ruin him, but his use of opioids persisted, and he was continually plagued by destructive behaviors throughout his career. This is not a unique story, but it is preserved in the historical record because Peirce was an originator of the philosophical school known as pragmatism. He made unique contributions to theories of inference and statistical coherence as well as the philosophy of science. His status had no impact on his vulnerability to the deleterious effects of opioid addiction, though. In the twentieth century, William S. Burroughs began from a position of privilege at Harvard University, and he descended rapidly downward when he began to use opioids. He faced some forms of vulnerability throughout his life, but he also had enough cultural cachet to be able to depict his fall in an autobiographical novel, *Junky*, and regain the status of a renowned author and intellectual. Finally, Ken Kesey popularized drug culture for the Beat Generation, supporting the use of a great range of drugs. However, the romance of drug exploration faded fast when opioids entered the picture, as they do not lead to mind expansion—they leave the mind closed, obliterated, and vulnerable to a downward fall.

Focusing on these cases can be a helpful reminder, but it also makes it difficult to see the broader patterns of addiction in culture at large. They highlight individual struggles, and they make it clear that people who can find social support are often able to navigate their experience of addiction successfully. Vulnerability to addiction is a natural tendency that is present across a wide range of people. Some people who struggle with addiction are creative; others are not. Some of them are famous, but the majority are not. But all of them should be acknowledged because everyone who suffers deserves our compassion and understanding. Unfortunately, the stories that are preserved are often the stories of famous or well-known people.

Against this backdrop, let's consider the emergence of the most recent wave of addiction, beginning with the rise in heroin addiction in major metropolitan areas during the 1960s.[73] Throughout this period, the stigma of being the "wrong kind of addict" tracked race and class biases as well as structural differences in access to material resources. It would commonly take the form of claims about moral decrepitude. Patterns of cultural imagination are likely to have made it easy for some people to associate intravenous drug use with ethnic groups who inhabited specific neighborhoods in major metropolitan areas—including Jewish, Italian, Irish, African American, and Latin American communities. The people who lacked access to material and social support bore the brunt of the criticism for the emerging opioid crisis. At the same time, similar patterns of behavior were being glorified in musical and artistic contexts. Put somewhat differently, broad patterns of cultural bias yielded a situation where the entitlement to material, medical, and social support was preserved for some groups of people, while bias was fostered against many others.[74]

The Velvet Underground sang about how shooting up could be an escape, because when "the smack begins to flow [you] really don't care anymore."[75] Given their cultural context, this makes sense. At the time, young users in New York City skewed toward the injection of heroin, perhaps because this was what was readily available. At the same time, the portrayal of this practice in art entrenched drug-based imagery by highlighting the immediate "rush" of shooting up. Intravenous injection provides a direct pathway to the brain, with almost immediate effects. It can quell anxiety and dysphoria, yield a momentary sense that things are okay, and support a nod into oblivion. Injecting heroin can also produce a powerful euphoric sense of warmth and a dropping out of the world into something like a dreaming state. This is what became a core feature of the cultural imagination. For many people, the anticipation of euphoria and the desire to experience dropping out might have been able to override the initial experience of nausea, itching, or discomfort. But as we know, the tragedy of withdrawal often lies in the future. As The Beatles sang, you often get to a point where you "need a fix 'cause [you're] going down."[76] Outside of the artistic contexts, the cycle of use often persists, with the avoidance of withdrawal becoming a primary motivation.

Throughout the 1960s, the delusion that happiness could be found in the rush came to prominence, and the experience of blunted sensibilities and oblivion saturated artistic contexts. We see the effects of this process in the many famous life stories that have intersected with opioids and ended with tragic overdoses from the 1960s through the present—for example, Billie Holiday, Lenny Bruce, Charlie Parker, Janis Joplin, Cory Monteith, and Philip Hoffman Skaggs. Again, these are the names that we remember. But the litany of lost lives always includes untold numbers of forgotten victims, and the absence of these stories from familiar narratives should be staggering upon reflection.

With all of these factors in mind, the main thing that we want to note is that the playing field is not level and it has never been. Some people receive care, support, or tools for navigating the difficulties of opioid addiction. For those who are perceived as the "right kind of addict," management is often medical. Indeed, this must be the case, given the role of corporate interests in managing the use and abuse of opioids. Furthermore, law enforcement and the American Medical Association have sought increased accountability for prescribing physicians. By contrast, when minorities and poor people become addicted, they are more likely to be policed or ignored. For many people, the most likely outcomes of addiction are overdose, incarceration, or a life saturated by sickness and fear. Prison rarely helps in the long run. Sometimes, it postpones drug abuse, but it often fosters other dysfunctional ways of coping, including finding more creative ways to secure access to opioids. Even in the best cases, however, jails and prisons in the United States abound with threats and exposure to constant challenges to viability. Furthermore, incarceration is often an undeserved fall to the bottom, socially, materially, and spiritually. These forms of trauma are foisted unequally upon African Americans, Indigenous people, other minorities, and poor people.[77]

Many of the forces that lead to this distribution of help and punishment center on figuring out who is the right kind of addict. This is a bad thing. From our perspective, what matters more is providing people with the opportunity to learn new habits, find new ways of being in the world, and cultivate links to others. For many, these opportunities provide a pathway for recovery. However, escaping addiction is far more difficult in a culture obsessed with allures that trigger the memory of a high and enhances the allure of a drug by glamorizing it, even as *using* opioids gets less satisfying with each use.[78] These are obstacles that can be navigated with help and support. But when opportunities to navigate addiction are so unevenly distributed, addiction often becomes deadly where compassion and understanding are withheld. Of course, it can also become deadly where these virtues are in place. The broader issue turns on the necessity of providing better forms of aid and assistance to everyone.

There were approximately 106,699 overdose-related deaths in the United States in 2021.[79] This fact must be understood in terms of the psychosocial and existential factors that are associated with the use of opioids in this country. The deformation of pleasant experiences and pervasive desire to avoid discomfort can foster despair and lead to the abandonment of hope. Many stories have been outlined in personal, scientific, and historical texts about the use and abuse of opioids.[80] Personal narratives often describe different lifestyles and ways of surviving, changing, and dying. They make it clear that there are many roads to becoming addicted. In recent years, death from addiction to Oxycontin has become a familiar tragedy in many towns and cities. There is an increasing awareness that many regular people are suffering.

In our current context, many parents are forced to observe the lives of their children being defiled by the wanton search for opioids. This happens across patterns of economic advantages and disadvantages. But people who are socially and economically oppressed and marginalized sometimes find that they must do this from a distance, as their children are jailed or experiencing other forms of harassment as the result of unequal access to treatment. These contexts rarely evoke the same forms of compassion that arise when addiction follows from *legitimized* uses of pain medications. In these cases and many others, biases often shape theorizing about the vulnerability to addiction. Sometimes, this leads to approaches that focus on various environmental as opposed to internal factors; other times, it leads people to do the opposite.[81] Far too often, though, the social understanding of addiction makes it difficult to see how patterns of addiction are shaped by the interactions between physiological and social factors. So, where should we go from here? Perhaps we can begin by thinking critically about the management of pain and opportunities for care and recovery.

5

MANAGING PAIN

I n 2015, Travis Rieder was involved in a motorcycle accident. His foot was crushed, and over the next year, he would have six surgeries attempting to reconstruct it. The complex experience of different kinds of pain—from the injury and the surgeries— was unlike anything he had experienced.[1] For a few weeks after the accident, he received morphine, fentanyl, or hydromorphone intravenously every four hours. As the need to manage diverse pains increased, this was supplemented with immediate-release Oxycodone and extended-release OxyContin. When he finally left the hospital, Rieder used the opioids he was prescribed responsibly. He attempted to taper off his usage in consultation with various medical professionals, but the physicians that he consulted disagreed about the right dose, and none of them actu- ally had a plan to help him taper his opioid use. As he tried to get off the pills, he began to experience classic forms of withdrawal, which were amplified by increasing experiences of anxiety.[2] He tried to get help, but this became difficult once it became clear that he was addicted—few doctors would talk to him, and the doctors who did told him to go to a methadone clinic.

Rieder recovered from his experience of dependence, and he describes his path to doing so in his book, *In Pain: A Bioethicist's*

Personal Struggle with Opioids.[3] Rieder had strong support networks in his personal as well as his professional life. He had family and friends who paid attention. As a bioethics professor at Johns Hopkins, he also understood the complexities of medical decision making, and he understood the factors that had produced his situation. Humans are powerless, though, when opioids bend various cognitive, affective, and pain-regulating processes. Social support and knowledge might be necessary to navigate addiction successfully, but they are certainly not sufficient. Rieder was lucky, and his experience reveals some of the reasons why people struggle to navigate pain management and addiction in the context of the U.S. medical system. Opioids are cheap, and this really matters because people are often reduced to insurance numbers. Furthermore, many medical professionals lack the knowledge necessary to help or lack the incentive to see patients as people with complex needs and experiences. We don't want to oversell this point. Since pain is an important facet of patient well-being, doctors must often prescribe drugs to treat it. The drive to minimize the pain that patients feel makes sense, but it doesn't prevent catastrophe, nor does it excuse the decisions that are often made in the context of pain management. Our aim in this chapter is not to offer an exhaustive account of these phenomena; we simply want to provide a brief overview of some of these factors and their impact on the current opioid epidemic.

PHARMACEUTICAL COMPANIES AND THE ADDICTION EPIDEMIC

Sometimes, experiences of pain can make survival seem almost impossible. The endogenous opioids evolved to help us navigate pain by mobilizing behavior and shaping neural activity that can support the pursuit of viability and eventual recovery. Ingested opioids also offer an escape from pain. However, they do so by minimizing connections to the complexities of life. They also produce a life that inevitably includes a lot of pain. The physiology of pain and the way that opioids manage pain entails that they become less effective over time. As sensitivity to pain—both physical and psychological—increases over time, this often creates a vicious cycle of need and use and a life where avoiding pain is the only goal that feels worthy of attention. Opioids are addictive, and there is no way of using them that avoids the risk of addiction. This is a real problem since opioids are neither difficult nor expensive to produce, and the demand for them has become ravenous. But this wasn't always the case.

The dangers of opioid abuse have been known for quite some time, but the introduction of OxyContin and the way that it was marketed produced a new class of challenges. Though there have been many participants in this crisis, the Sackler family has probably earned the most infamy. Their money and their name have long been associated with art, culture, and society—but they have also become a cautionary tale. Arthur Sackler was involved in marketing Valium, an addictive antianxiety medication. An aggressive marketing campaign, beginning in the 1960s, suggested that Valium could be used quite broadly with little risk of addiction. In a world increasingly saturated by anxiety, this helped Valium to become one of the most profitable pharmaceuticals ever produced. Something similar happened with Oxy-Contin and other opioid drugs. Beginning in the 1990s, Purdue Pharma—which was owned by the Sackler family—initiated marketing campaigns that presented misleading information about the addictiveness of OxyContin to doctors and consumers, leading to high levels of overprescription.[4] Purdue Pharma continually claimed that OxyContin was not addictive, making it possible to imagine it as a safe and cost-effective form of pain relief.[5] Even as they learned how addictive these drugs were, they continued to market them as *less addictive* than other forms of pain relief. Furthermore, while they are best known for manu-facturing the opioids that produced the current opioid epidemic, they also manufactured Naloxone—the common antidote for overdose. So, they profited from both use and overuse.[6]

All of this matters because many doctors found themselves needing to see patients as quickly as possible to support chang-ing profit models. This led to an increased focus on quick fixes, fewer opportunities for genuine interaction, and reduced motivation to address the causes of pain. Finally, interactions between insurance companies and pharmaceutical companies helped to incentivize the use of relatively cheap opioids while making it more difficult to pursue other pain management strat-egies (including physical and cognitive therapies, as well as more expensive pain medications like intravenous acetaminophen). This was a perfect storm of interacting factors, which contrib-uted to a sixfold increase in the death rate from opioids between 1999–2017.[7]

Opioids can treat pain, they can support incredible profit margins, and they can leave behind an uncounted human toll. In the cutthroat world of pharmaceutical manufacturing and marketing, cashing in on opioids was one way to keep compa-nies profitable, but the massive advertising campaigns that led to high levels of overprescription were not simply par for the course. Pharmaceutical companies spend a substantial percent-age of their budget on marketing. In the case of Purdue Pharma, this included paying sales representatives to travel the country, meet with doctors, and promote their latest products. It included providing gifts in the name of advertising and creating perverse incentives for prescribing drugs like OxyContin.[8] Additionally, it included steering doctors toward opioids as solutions for diverse problems, while downplaying the risk of addiction.[9] It also included television advertising and other forms of media that directly targeted consumers. While pharmaceutical companies started the fire, numerous other actors played supporting roles in supplying increasing numbers of people with opioids.

The corporate imperative to grow at all costs led Purdue Pharma to lie about the risks of the drugs they were creating and profit from the creation of a crisis and attempts to stop it. Approximately $31 billion was made by selling OxyContin, which was enough to buy catastrophic indifference. Numerous studies have confirmed that doctors are influenced by marketing from pharmaceutical representatives.[10] Like the rest of us, they are vulnerable to deceptive marketing. Of course, most doctors are not just in it for the money. They want to do good, but the healthcare system in the United States does not always reward those who pursue good outcomes. Sometimes, it rewards those who avoid difficult questions and difficult solutions. It would be easy to assume that since doctors aren't superhuman healers, they are little more than high-status drug dealers.[11] However, while many doctors did prescribe far more opioids than were necessary, the current crisis could not have developed without doctors who had good intentions, never ran pill mills, and empathetically prescribed the pills that made so many people dope-sick.

It is not easy to determine who is vulnerable to opioid abuse or addiction. Even if it were, few doctors would have the time to figure out which patients are vulnerable. We know that life histories and broader proclivities toward addiction impact the risk of opioid addiction.[12] In an ideal world, doctors would pursue an integrative approach to pain management, which centers the challenges, vulnerabilities, and psychosocial issues that shape the likelihood of addiction.[13] Doctors have little time to engage with patients directly, though, or to keep up with results in pharmaceutical science. So, many patients find that they never develop real relationships with their doctors, and doctors experience high levels of burnout. Even when doctors have good intentions, persisting shortages of doctors, the demand for high patient turnover (to sustain profit), and other constraints on modern medical systems sometimes yield inadequate care.

PRESCRIBING OPIOIDS

To be clear, most doctors want to listen to patients. Ideally, most of them would be happy to provide treatments without the consequences of opioid addiction, but opioids are an important tool in the arsenal for treating pain. While they are risky, advocating for the abolition of medical opioids would be misguided, as they can be the best—and sometimes the only—viable option for treating pain. Furthermore, there are medical interventions with extremely painful side effects. Here, too, doctors want to manage pain, but they must find ways to couple painful therapies with drugs that will ameliorate pain. This was true with Rieder's surgeries, and it is true of many cancer treatments. It should come as no surprise, then, that Sloan Kettering, a cancer hospital, was an early leader in opioid prescriptions. Patients were suffering. Patient organizations were demanding drugs that would help, even if there was a risk of addiction, and physicians, such as Kathleen Foley, wanted to help manage pain—for her, this meant prescribing opioids.[14]

It is important to remember that many doctors were initially ignorant about the risks of using opioids and that OxyContin was initially advertised as being less addictive than other opioids.

So, it was easy to prescribe them to people who could be treated in other ways. Once the habit of prescribing opioids for every workplace and sports injury became stabilized, it would become hard to break. Against this backdrop, the problem of aggressive advertising was especially acute. In the context of the American medical system, cost is king. Treating the symptoms is often cheaper and easier than addressing an underlying problem. Doctors in the American system often try the cheapest and fastest option first—they must, given the ways that the economic burden of medical treatment is distributed to patients and insurance companies. OxyContin was just far cheaper to prescribe than other therapies that might permanently reduce pain.

Of course, there are guidelines in place for prescribing opioids for pain management.[15] However, the authoritative sources offer conflicting advice, and their recommendations sometimes contradict longstanding practices.[16] Confusing advice about how to change made it rational for doctors to hang on to past courses of action. This allowed overprescription to continue even after it was clear that OxyContin was highly addictive. Continuing education might help some doctors to consider alternative pain treatments. Moreover, clear guidelines for prescribing opioids—along with mechanisms that sustain accountability and transparency—might reduce the likelihood that opioids will be pursued as a first option. There is a continued need for better access to alternatives and insurance coverage that can support the pursuit of alternative treatments.[17] However, unless patients can access and afford other options, it won't really matter whether doctors consider them or not.

There have also been doctors who rationalized opioid prescriptions, thinking that patients would get the drugs from someone else if they did not comply. Internet resources made it easier for people seeking opioids to find doctors who would prescribe them. Along with direct drug advertising, this seems to have created a class of "professional patients" who were taking advantage of "compassionate doctors" in ways that created the epidemic of addiction.[18] Some doctors have even confessed to knowing about and ignoring the risks of opioids. The American medical system made it easy for them to make money with opioids, and some doctors gave in to this temptation, relying on "cyber-pharmacies" to guarantee that their patients would have access to the drugs that they were prescribing.[19]

As the opioid epidemic spread, whistles were blown on unscrupulous doctors who were prescribing opioids in excess of the reasonable needs of their patients. These doctors were running the pill mills that fanned the flames of addiction in their communities. When the DEA finally cracked down on prescription opioids, it was too late to help many people. Some unscrupulous doctors continued to provide access to opioids even after government crackdowns began. The persistence of perverse incentives, along with relatively lax attempts at regulation, made it far too easy for doctors to ignore the dangers of opioids until it was too late for too many people. Just as importantly, getting off opioids is hard, and there is little guidance or support for doing so. So, many individuals began to turn to illicit markets to retain access to the opioids they needed. Thus, heroin and fentanyl became readily available in communities that once had high rates of opioid prescriptions, and markets for them began to appear in communities where they had never been seen before. So, the cycle of profit continues, even if it does so under a different guise.

EDUCATION AND TRANSPARENCY

This isn't a new story. The opioid epidemic has targeted vulnerable people, and the difficult issue for healthcare providers is to find ways of managing pain that don't *systematically* disadvantage any particular group of people. This is far easier said than done. Many societies have tried to stop opioid abuse, and many have been overwhelmed by the force of profit seeking. This happened with the Opium Wars, where the British fought to promote opium in China and to stop China from protecting its citizens. In this case, the desire for profit won because it was supported by an enormous amount of economic and state power. But there may be ways for state power to pull in the other direction. For example, increased transparency might serve as one small way of fighting back against pharmaceutical interests.[20] This is possible because records of prescriptions are becoming available, yielding more transparency than ever before.[21] Of course, preserving transparency requires doctors to spend more time documenting work and less time helping patients. For such transparency to make a difference, people must be made aware of the available resources, and they must be willing and able to *read* the published records. It is hard to know if this is possible in our current context, as most local newspapers have either shut down or cut reporters, most news channels have become sites of political contestation, and social media continually distorts the epistemic landscape.

This situation is also complicated by the fact that patients must often advocate for their own treatment within the American medical system. This is a tricky situation, as patients often know what is wrong, but they don't always know how to fix it. So, sometimes patients want access to opioids, and sometimes they pressure doctors to prescribe them.[22] This is often a matter of self-advocacy. Since doctors typically want to help their patients, many of them empathetically accept such requests. This is part of what allowed OxyContin to play into the desire for a magic bullet—a pill that could fix problems without significant effort and the cost of providing long-term social support for patients.

But desires for pain relief persist, even with new medicines. So, it is important to remember that doctors are human, and many will be tempted to overprescribe unless they have good reasons not to. This is why things like regulation, oversight, and transparency are so important.[23] The public must also be able to know who is overprescribing, and they must care enough to take advantage of this information.

To be clear, transparency can only ever be one part of the story, and it has significant limitations. Any regulatory regime can be evaded if the incentives are right.[24] Where billions of dollars can be made by bending or breaking the rules, many people will attempt to do so. This is the situation we find ourselves in, and it makes it difficult to rebuild the trust that Americans once had in the medical profession. It might be possible, but it will take a lot of time and effort. After all, the opioid epidemic has laid bare numerous challenges that arise when doctors take the easy path instead of the right one. For example, paternalistic attitudes shaped the willingness to prescribe opioids. Often, doctors attempted to ameliorate complaints by women with opioids.[25] This wasn't driven by a willingness to address problems, and ameliorating complaints without addressing the underlying cause of pain led to high levels of overprescription for women and high numbers of women who became addicted to opioids. By contrast, many doctors appear to assume that Black patients feel less pain than white patients, so they prescribe less pain medication for the same ailments.[26] This yields unequal access to pain relief and further reductions in trust in the healthcare system. The long and short of this is that numerous social ills surface in the treatment of pain, and these must be addressed for people to once again trust the healthcare system.

In moving forward, accountability practices might be established for individuals and institutions that attempt to evade governmental regulations. Moreover, there might be ways to reduce the temptation and willingness to sacrifice public health for profit. It is unclear whether the deep structural changes that would achieve these ends are actually possible. The Hippocratic oath calls on doctors to do no harm, but this requires knowing what causes harm and what does not. Increased transparency can play an important role in this process. It can help doctors see what works and see where things are worth the extra cost—as such, it can help them develop better treatment strategies for pain and addiction.[27] The urge to understand pain management has led to an expanding culture of medical experimentation as well as important explorations of the consequences of medical decision making. Doctors must keep learning if they want to provide treatments that are not just familiar but useful.

Most doctors want to see and engage with their patients. This might become difficult as regulatory burdens grow. If the aim is helping patients and placating pain without causing harm through bias or accidentally leading people into addiction, then the drive to limit access to opioids must be balanced against the recognition that equal access to pain treatment is a basic human right. This will require training programs that can keep track of changes in the world—such as the growing use of fentanyl—adjusting and adapting as necessary. It will also require integrating ongoing research, which is exploring ways to treat pain that are nonaddictive and pursuing knowledge of ways to use opioids that minimize the risk of addiction.[28] At present, however, pain management is a medical specialty that is largely disconnected from the specialty of treating addiction. As a result, many doctors will be unlikely to appreciate all of the potential consequences of their prescription decisions. Of course, few people think that addiction is a rare outcome of the prescription of opioids for medical purposes.[29] Many doctors now have addiction on their minds when they prescribe opioids, but few know everything that they need to know about opioids when they graduate from medical school. While they can be taught, improved medical education is necessary for doctors to avoid the most problematic outcomes of prescribing powerful pain medications.

To be clear, education in pain management is a constant and recurring need that impacts all medical specialties, from cancer care to sports injuries and COVID-19 recovery. Much ongoing education focuses on new procedures, but many doctors must also be trained in pain management, the use of opioids, and alternative pain management strategies. These include knowledge of new painkillers and how they work, knowledge of attention training practices, new research on biofeedback and far infrared light therapy, various kinds of topical treatment, and various forms of physical and movement therapies. Eventually, some of these research findings will begin to shape public health policies, but this is something that will require a concerted effort to understand the latest research by a diverse range of people in a diverse range of social positions.

There is no escape from pain. Of course, different people experience different kinds of pain. You might experience injuries, childbirth, toothaches, arthritis, grief and loss, or one of the countless other forms of pain that saturate our world. You might even experience pains that are disconnected from external stimuli, such as the pains in phantom limbs that are common after amputation.[30] Pain is also a totalizing experience. Such experiences, especially when they take the form of chronic pain, are things we would like to avoid, but doing so isn't always easy. Perhaps this is why the assumption that pain is a problem that can be solved gave way to the current epidemic of opioid addiction, which has killed more Americans than many of the wars we have taken part in. Fortunately, many doctors and researchers are searching for alternatives and acknowledging that a one-size-fits-all approach is highly unlikely. Science has helped to create this problem, and hopefully, science will be able to help us to solve it.

In everyday thought, expressions of compassion are often associated with medical professionals. For example, we might think about the heroic compassion of historical figures like Florence Nightingale and the other nurses who worked to relieve suffering during the Crimean War. They recognized that easing pain was an important aspect of preserving the autonomy and well-being of those who had been wounded—and that was what mattered to them. Preserving autonomy and well-being, to the extent that doing so is possible, is a normative goal and a human aspiration, even as we approach death. Modern medicine has the potential to increase autonomy and support the cultivation of human well-being, but this is another incredibly complex situation.

Consider the fact that death is often painful. For those who are fortunate, hospice care can help the dying to be treated like people—but there are many who do not receive this kind of palliative care. Instead, they are given opioids to help them escape from pains that can't be treated easily and to put experiential distance between them and terminal diseases. Many hope to die at home, surrounded by the people and the things that made their lives meaningful. For those who cannot be cared for at home, hospices allow the living to approach death without the fear that accompanies the hospital environment. Far too often, life ends with a person being forgotten amidst the swarm of charts and white coats. This is not what people deserve.

Modern medicine has contributed to diminishing autonomy and well-being in a wide range of contexts. The approach to managing opioid addiction is another context where this has occurred. Indeed, participants in medical experiments that never should have been carried out have also needed to deal with the resulting patterns of substance abuse.[31] How can we guard against this? To begin with, expressions of pain are a social signal. Most humans are attentive to the bodily and facial expressions that signal physical and psychological pains. Since we are social creatures, a drive to care for those who are suffering is an evolutionary advantage.[32] When someone is in pain, we know that they need help. We might not always offer that help, but pain is instantly recognizable. Its prevalence and universality can help us to relate to others. Of course, most of us privilege our own pain and the pains of those who we see as members of our own groups—as Buddhist philosophers have often noted, and in our current social context, this means that pain management decisions frequently reflect and perpetuate racial injustice.[33] However, if we cultivate a heightened receptivity to the signal, the expressions of pain can capture our attention and help us understand the challenges that others are facing. This is a form of compassion that holds our social world together, orienting us toward socially meaningful forms of thought and action.[34]

A disposition toward compassion might be an evolved trait, but compassion is also a virtue and a habit of social engagement that must be cultivated or allowed to atrophy. For Adam Smith, compassion and other moral sentiments were relations we have with the world.[35] For many Buddhist philosophers, compassion was the piece of an ideal human orientation that should be cultivated in pursuit of individual and collective liberation. Whichever path we follow, compassion is important for each of us and for our well-being, so it is also in our self-interest to pursue it. Of course, we also need to figure out how to help people manage their pains. As we noted in an earlier chapter, the neural basis of pain is complex, engaging diverse cortical and subcortical regions. Luckily, this also means that there are many ways to address pain. Just as a brain can register pain in an amputated limb, pains can also be mollified with placebos.

Placebos can activate the same neural regions that are activated by opioids (figure 5.1).[36] The placebo effect is not just a figment of our imagination—it is a brain-based response that tends to be most effective in the short term in ways that appear to be tied to our expectations.[37] The expectation of pain relief appears to stimulate the brain in ways that reduce pain perception. Intriguingly, placebos appear to activate the u-opioid receptors.[38] Perhaps this reflects the evolutionary advantage of allowing the brain to produce feelings of well-being merely as a function of expectations.[39] The initial response to a placebo does not last, however, if pain has a robust physical cause.

The placebo effect in pain management is well-studied.[40] The facts that expectations can exacerbate pain and that stress-related events can make self-reports of pain more severe have also been

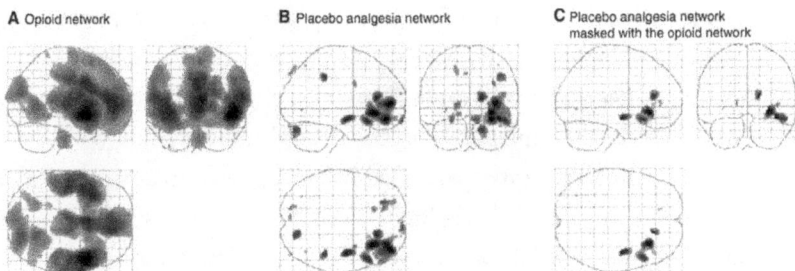

FIGURE 5.1 Regions of the brain activated by (A) opioids (specifically, remifentanil) and (B) a placebo; (C) depicts the pattern of overlap between opioid stimulation and placebo conditions. Reprinted with permission from the American Association for the Advancement of Science from Petrovic et al. 2002, 1738.

researched.[41] The brain is extraordinarily complex, and we still do not have a complete picture of all the ways that aspects of our well-being interact. That said, these effects point to the fact that we should treat the brain with more than just chemicals. Just as a brain can help ease pain without the application of opioids, it can also make pain more severe. This effect can be visualized by examining epigenetic changes in brain cells and by observing maladaptive changes in supraspinal reorganization.[42] We can also see this in the way that meditation practices can shape attention in ways that can sometimes be almost as effective as painkillers. Open monitoring meditation, for example, seems to provide a strategy for reducing the unpleasantness of pain even while the experience of pain remains intense.[43] Pursuing this approach is not a quick fix, however. It requires a lot of work. While it does seem to carry far fewer side effects, even meditation can sometimes yield aversive experiences.[44] None of this, however, compares to the way that opioids can help or hurt us all.

The first order of business is saving lives, minimizing harm, and offering opportunities for success. This requires acknowledging that some people are more vulnerable to addiction by virtue of their families, histories of stress, and perhaps even facts about their neurobiology.[45] As Nancy Campbell reminds us, however, a *critical neuroscience of addiction* must situate substance use and abuse in the context of the larger social milieu.[46] Pain does not occur in a vacuum. It is shaped by stress, social trauma, and attempts to manage various kinds of challenges. Just as importantly, there are social contexts where people are hurt and where they decide to hurt themselves. There are also pains that unfold in the context of pursuing opioids—some physical, social, or psychological. We need to provide people with support and resources for grappling with the pains that unfold in the context of addiction. We also need support structures that target both the physical and the psychological aspects of pain in ways that support successful recovery from addiction. We need better education for doctors, and more research into alternative techniques for pain management. We need better insurance coverage for alternative treatments and better ways of treating the underlying causes of diverse pains. This is not something that can be achieved with better drugs. This situation demands working on the causes of pain and trying to change them rather than mask them.

6

ACCOUNTABILITY AND
REDUCING HARM

ragedy is a pervasive feature of human life. The Greek tragedies highlighted the terrible ends that are produced by the confluence of personal failings and impossible situations. Likewise, the central narrative of the Sanskrit epic, the Mahābhārata, detailed the loss, tragedy, and disquietude that unfold at the intersection between a gambling addiction and an unbounded craving for power. Drugs—including narcotics—have had an enormous impact on human history, and addiction to opioids is also ripe for tragedy.[1] As with the tragedies in these classical texts, we should avoid nihilistic or romantic assumptions about freedom and agency, and we should ask: What does it take to create possibilities for hope and redemption in the face of tragedy?

As we have noted at many points throughout this book, opioids are primarily a way to avoid or deaden pain and to reduce connections to a world that is shaped by suffering and discomfort. This is why Marx famously referred to organized religion as an opium for the masses; for him, religion was a way of escaping from feelings that could drive creative action and attempts to transform the world. To learn from tragedy and loss, we must fight against the impulse to run from the pain. This is hard, given that tragedy will always pervade the world of opioid use and abuse. A life governed by addiction might be shaped by exhaustion, ostracism, degradation, or shame. It might include experiences of incarceration, hospitals, methadone clinics, or therapeutic communities. It might support actions that only seem rational in light of a pervasive need to pursue opioids and avoid pain. Reading about such tragedies could support a partial understanding of the patterns of loss and unfulfilled dreams that opioids create every day. Losing a friend or family member to opioids might evoke a deeply visceral response (as it did when Jay's friend Billy Toth died at seventeen). But we need to think more creatively and transformatively about how to navigate such experiences.

Addiction is a form of suffering, and it will remain so whether we assume that the ultimate source of addiction is the brain, society, or the interactions among brains, bodies, and worlds.[2] In suffering, the entire range of human possibility is on display. On the one hand, the decision to try opioids might be a choice. On the other hand, opioid addiction restricts but does not eliminate choices. Likewise, opioids take many lives—and dying senselessly is always a tragedy, but those who witness such tragedies are also likely to feel their own forms of existential grief. Our aim in this final chapter is to explore the possibility of pursuing freedom and agency in a world that has been shaped by the tragedy of our current opioid crisis. We emphasize decriminalization where possible, we argue that a culture of fairness should be promoted, and we emphasize the development of lifeways that are favorable for the pursuit of more humane possibilities. To that end, it will help to begin by saying a bit about how we ended up in a situation where the pursuit of more humane possibilities is something that needs to be argued for.

THE SOCIAL CONSTRUCTION OF "THE AMERICAN JUNKIE"

The "American junkie" was created by a process of cultural production that was guided by assumptions about moral virtue, and biases related to race, ethnicity, and class.[3] During the first half of the twentieth century, it became common for the social and political elite to argue that people who used cocaine, opioids, or alcohol were morally vicious and that they would destroy the moral fabric of society. The Harrison Act of 1914 was the first law to ban the nonmedical use of opioids and cocaine. It also imposed strict regulations on producing, importing, and distributing these substances. The Volstead Act of 1919 then established new law enforcement agencies to enforce the Harrison Act. By the 1930s, a Federal Bureau of Narcotics was established to find and punish those who were involved in the illicit use and abuse of opioids and other substances. The changes in federal practices were only part of the story, though. Alongside increases in policing and incarceration rates, public attitudes had started to shift. People started to treat addiction as a moral failing, and they began to point toward the deficiencies of the poor, Black, and immigrant communities where addiction seemed to be occurring. At this point, incarceration became one of the most common outcomes of addiction. Indeed, by the mid-1930s, nearly half of the people who were incarcerated in the United States had been convicted of drug offenses.

There were, however, attempts to resist this dominant narrative. The U.S. Narcotic Farm, which opened its doors in 1935, was a state-sponsored treatment facility for addiction.[4] It wasn't exactly a prison, nor was it really an asylum. The patients at the Narcotic Farm were diverse with respect to race, class, and gender. Some of them were artists. Sonny Rollins and Chet Baker were patients there for a time, as was William S. Burroughs, who wrote about his experience in *Junky*. Some of the patients had been convicted of drug-related crimes, but others were seeking rehabilitation of their own accord. At its core, the Narcotic Farm was an attempt to deal with addiction in ways that didn't depend upon the worst aspects of the criminal justice system.

Lawrence Kolb, a psychiatrist, was the first head of the Narcotic Farm. He thought that the willingness to pursue incarceration reflected ignorance about the cause of addiction. More specifically, he held that the most troubling forms of addiction should be understood as medical disorders, which were reflected in personality traits.[5] Kolb would spend much of his life as a spokesman for the medical treatment of opioid addiction. He wanted to make sure that people were not being punished for being mentally ill, and he held that people who were addicted to opioids could be healed and become productive members of society. This latter assumption was a guiding principle of the therapeutic model at the Narcotic Farm.

In discussing the origin of this facility, Nancy Campbell quotes President Franklin Roosevelt, describing the Narcotic Farm as an institution dedicated "to the noble purposes of restoring our fellow men from the abject slavery of the narcotic habit" and a place where "victims of the opium habit will be restored to usefulness."[6] This is why the Narcotic Farm was located in a beautiful part of Kentucky, surrounded by horses, mountains, and fresh air (figure 6.1). Its core therapeutic values centered on the pursuit of moral improvement. This often meant making patients engage in hard work—including agricultural labor on the farm and making furniture that would be sent to prisons—but it also included time for both recreation and intellectual development.[7] At this high level of abstraction, it almost feels like this should have been an ideal situation for the management

FIGURE 6.1 The Narcotic Farm, later known as the United States Public Health Service Hospital, Lexington, Kentucky. The National Library of Medicine, digital collections. Public domain image.

of addiction and the process of recovery. But relapse rates were consistently high at the Narcotic Farm.

The explanation for the failure of this model would take us far afield from the story that we want to tell. For our purposes, the key thing to note is that the Narcotic Farm was also a research laboratory, and many of the physicians there were interested in finding better ways to manage withdrawal. In collaboration with the National Institute of Drug Abuse, they developed drugs like nalorphine, which they hoped would be less addictive than morphine and heroin.[8] But this situation was complicated by two factors: first, the patients continued to be seen as mentally ill, morally deficient, or less than full humans. Second, this research was carried out prior to the entrenchment of Institutional Review Boards—so there were few regulatory structures in place to ensure that patients were treated ethically. As a result, risk assessments were often minimal, and patients were subjected to morally dubious experiments, which employed everything from radiation to psychotropic drugs.[9] The Narcotic Farm thus became an ideal setting for unethical research. As rehabilitation faded into the background, the core aim became finding better ways to manage opioid addiction medically.

A final plank in our current strategies for regulating opioid addiction was put in place when Richard Nixon declared a war on drugs in 1971. This war on drugs would end up including the use of military force, focusing on places where the drugs were being produced—and since war is always the outer end of state-craft, this made room for greed, ignorance, and even criminal-ity to saturate it. There was nothing new about this situation. Wars have often been fought for opium and against opium, and after each one of these wars, part of the fallout has been a rise in opium use and abuse.[10] But in this context, war was never going to be the right metaphor.

Wars occur between people, and the primary casualties of the war on drugs have always been the people who use and become addicted to drugs. Opioids have been a part of human society for thousands of years, and their use persists in the face of these wars, even if there are moments where use rises or falls. But enforcing laws surrounding the sale, production, and use of opioids does more than simply combat drug use; it always entails that people are killed, incarcerated, or marginalized for becoming addicted to opioids. Like all wars, this one has always had substantially broad social costs.

MOTHERHOOD AND ADDICTION:
THE PERSISTENCE OF MORALISM

In our current context, opioids remain a valuable resource. They might continue to be so even if chemists find ways to minimize the addictive aspects of these substances while retaining their medicinal uses. At this point, however, we cannot be sure. So, our aim should be minimizing harm. Fortunately, while we continue to be vulnerable to addiction in the same way people have always been, we also have the benefit of hindsight. Epidemics of opioid abuse have crested and fallen in the past, and paying attention to that history might help us to figure out what works and what does not. In some respects, the "damned opium mess" that was described by Charles Henry Brent in the early twentieth century is still with us.[11] Indeed, we are living through the greatest crisis of addiction in living memory, and this means that we all need to find better ways of dealing with this situation.[12]

Dupont used to advertise "better living through chemistry." This is probably exactly what many scientists and physicians assumed they were providing. Some of the new molecules were more effective in fighting pain, but they were also more addictive and more likely to lead to overdoses. Oxycodone, for example, has been very effective in getting people addicted to opioids. Of course, there are substances that have helped people to find better paths, including methadone and buprenorphine. But the imperative to design safer painkillers that decrease vulnerability to overdose, abuse, and addiction is a tall order.[13] Mistakes along the way are likely to generate new problems that must be dealt with. We cannot overstate this fact. From 1993 to 2010, the rate of hospitalization for opioid overdoses in the United States increased from 2.4 per 100,000 persons to 15.9 per 100,000 persons. This is frightening, but during this same period, the largest increase in opioid use—over 1,000 percent—was recorded among pregnant women. Overdoses among pregnant women—mainly from fentanyl—skyrocketed over this period as well.[14]

Some state legislatures are debating the possibility of jailing women who have a baby while they are addicted to OxyContin. In many contexts, these are the same states that want to impose strict regulations on access to abortion. Those who support such practices claim that a baby's rights are being violated if they are born facing addiction, but they say nothing about what will happen to a baby whose mother is immediately incarcerated. Such policies and others like them are morally abhorrent.[15] The public outrage at such policies cuts across party lines and differences in class, race, and religion. Even on the assumption that addicted mothers should be held morally responsible, incarceration is unlikely to fix any relevant problem. Indeed, it is likely to cause more problems. But before we consider this, there are numerous other factors that we must take into account.

We know all too well that there are conditions that make people especially vulnerable to addiction and relapse. These include depression, chronic stress, and chronic pain. Across numerous contexts, women are more likely to suffer from depression. There may be biological differences at play, but far more importantly, social conditions, ranging from access to material resources to patterns of intimate partner violence and diverse forms of gender-related biases and violence are all at play as well.[16] There are also critical justice issues that arise when attempting to manage opioid dependency during pregnancy or immediately after a baby is born.[17] These include questions about the use of methadone or buprenorphine during pregnancy or while lactating.[18] They also included the fact that some babies who are born to addicted mothers are born prematurely, have a lower birth weight, and suffer from malnutrition. However, these are issues that should be addressed directly, not through a lens that centers on the assumption that a mother is "the wrong kind of addict." In many cases, we should begin with the hypothesis that these patterns reflect racial and class-based differences in social stress, as well as differential patterns of access to material resources.[19]

In many cases, being born to an addicted mother does have consequences later in life, but it is nearly impossible to determine which of these effects have been caused by race- and class-based variables.[20] Put much too simply, we don't know if exposure to opioids in utero increases later vulnerability to addiction. We also don't know how resilient and adaptable an infant can be if they have the support of a caregiver who can help them manage the challenges they face. The existing evidence has been shaped by patterns of biased data collection and interpretation, and it fails to take the complexity of physiological systems into account.[21] That said, we should not deny that an infant will suffer if they experience withdrawal the moment that they enter the world. Indeed, this is one of the main issues that must be addressed. There is some reason to believe that buprenorphine can be used to help infants who experience withdrawal.[22] Far more importantly, however, we believe that human contact, as well as support from a caregiver, are likely to play critical roles in managing this distressing experience—and this is to say nothing of the significance of care and support over the course of development.

Our key point here is that we should center harm reduction. To do so, we must invest in therapeutic and educational interventions that prevent things from reaching this point rather than doubling down on strategies that cause further harm. Policies that lead to the incarceration of addicted mothers ignore the complexities of navigating addiction while pregnant. They also ignore the fact that having a baby is often the primary reason why someone seeks treatment for addiction. Addressing these issues will require moving beyond the moral disgust that fills the public space and beyond the experiences of grief regarding a life lost. There is no doubt that such expressions are warranted, and they highlight the outrageous loss of life and the problems with a system that has created vulnerability for abuse in the first place. However, they do not offer a plausible strategy for navigating the constraints on agency that are revealed in these contexts.

HARM REDUCTION

Perhaps the way to think about such issues is to begin by recognizing that people who are addicted to opioids have often been thought of as disposable. At times, they have been murdered by law enforcement, but the slower and socially acceptable "solution" has often been letting people who are addicted to opioids die in prison. Within our culture, there is a drive to hold people accountable. Perhaps the legal and moral fabric of our culture necessitates these practices, but the current opioid epidemic reminds us that vulnerability to addiction is part of our evolutionary inheritance. Likewise, the observable patterns of deaths and ruined lives give us every reason to think that histories of trauma and conditions of economic and personal distress have a profound impact on who will become addicted to opioids. In her erudite book *Undoing Drugs: The Untold Story of Harm Reduction and the Future of Addiction*, Maia Szalavitz explores the way that "the uniquely moral nature of the way we treat addicts as both sick and criminal reinforces stigma."[23] She rightly notes that attempts to prevent people from using drugs rarely work, and attempts at policing drug use drive incarceration rates through the roof. She argues that we can focus attention on reducing the harm of addiction. This means acknowledging that people will engage in risky behavior but attempting to limit the harm they are likely to experience.

Many policy makers have started to get behind methods of rehabilitation and harm reduction. These policies often include things like needle exchanges or access to methadone or buprenorphine for maintenance.[24] At least in part, the shifting attitudes reflect the fact that our current opioid crisis was fueled by prescriptions for pain medications. Of course, the resulting demand inevitably led to the development of an underground market and nonmedical supply chains. This was partly a result of prescription opioids becoming harder to access and a blurring of the boundaries between recreational use, medical use, and addiction. There is a broad consensus that the current crisis occurred despite the best efforts of many people and that it reflects the tragic patterns of variation in the vulnerability to addiction.[25] So, there is a growing *public emphasis* on the use of therapy and medical treatment instead of relying on criminal enforcement. Yet spending on criminal enforcement continues to outstrip funding for every other social service in the United States.

This matters because people are not typically provided with the services they need in their communities. They are provided with limited forms of aid and support, and this contributes to ongoing support for policing and incarceration. This is the predictable result of the culture of regulation that came into view throughout the twentieth century. Where should we go from here? To begin with, we now know far more about the addicted brain than we ever have before.[26] Brains are part of a body, bodies have histories, and bodies are always embedded in societies that have histories of their own.[27] Sometimes, it is right to focus on the restoration of function, but rehabilitation must always focus on the whole person, not just the brain. To be clear, this process must often begin with medical interventions, and the first step toward recovery will often require moving the brain and the body away from habituated forms of dysregulation. It is critical to remember that these forms of dysregulation affect numerous features of embodied and social minds.

This is not a unique feature of addiction, but addiction and recovery reveal a commonly submerged fact that underlies all human cognition: we are a social species, and many of the traits that are typical of our species have evolved and typically develop in ways that can facilitate our deeply social capacities. This is one of the primary reasons why recovery often requires personal and social interventions that moderate the motivation to use or abuse opioids. However, in biology and ethics, it is often easier to focus on individuals. In both cases, ignoring the fact that individuals both shape and are shaped by their society leads us to miss the forest for the trees.[28]

Just as importantly, we must remember that the playing field has never been level in the context of opioid addiction. Factors like race, gender, and socioeconomic status impact the likelihood of receiving treatment instead of being sent to prison. These factors also impact the likelihood of having insurance or private resources to pay for assistance. Finally, these factors often affect where a person lives or works, the ongoing stresses in their lives, and the likelihood that they will be exposed to opioids. Most people who use opioids do not become addicted, but there are far too many who do. In the context of opioid addiction, pursuing justice means preventing addiction before it starts. This requires knowing more and knowing in ways that are sensitive to the individual and structural biases that have commonly shaped our collective understanding of addiction.

There are many understudied variations in how bodies process opioids. For example, there might be gender differences in vulnerability to addiction, which are tied to differences in the experience of pain.[29] Before we settle on this explanation, we must understand the biology of these differences and the ways that similar patterns occur in the absence of gender differences. Otherwise, we risk re-entrenching unfair and unequal access to strategies for managing pain. We also need to figure out who is vulnerable to addiction so that the pain experienced by people who are injured in accidents, playing sports, or while engaging in physical labor can be addressed without the risks of addiction. To some extent, personalized medicine might help by taking into account the unique risk factors of patients, but this opens the door to further forms of medical discrimination, which are likely to have a disproportionate impact on people of color, poor people, and women.[30] Finally, we need to recognize that addiction treatment is not completed quickly and painlessly. It can take years, and access to treatment needs to be available as long as it is required. This becomes even more important as we realize that relapse is the norm rather than the exception and that people can make it through relapses and still get back on track.

In reflecting on phenomena like these, Szalavitz emphasizes the diversity of treatments that might be necessary if we really want to reduce harm. For example, she highlights the ways that addiction is commonly entangled with depression in ways that suggest that SSRIs and related drugs might be helpful in supporting a sense of safety on the road to recovery.[31] She also emphasizes the importance of attending to various forms of neurodiversity, which shape the experience of addiction and patterns of addictive behavior.[32] Against this backdrop, we must remember that there are many different ways to pursue rehabilitation and habilitation.[33] Rehabilitation is a matter of returning to a previous baseline; habilitation is the process of acquiring or improving skills for flourishing in a socially complex world. To reduce harm, we must create safe places for acquiring and using opioids where this is necessary. We also need to find ways to decrease use, where possible, and to mitigate the harms of abuse. Finally, we need to find ways to increase recovery and human prospects. Often, these processes will begin by decreasing the likelihood of overdose.

Naloxone first appeared as a medicine to reverse opioid overdose in 1971.[34] Since then, several opioid antagonists have been introduced that help to reverse the breakdown of normal lung functioning in the context of an overdose. This includes the more potent antagonist, nalmephene. There are complex questions about how to allocate these resources in ways that will promote harm reduction. In New York in the 1950s and 1960s, overdoses were commonly assumed to be a problem faced mostly by Black people.[35] Indeed, at this point, deaths from overdose were most commonly observed among people of color. We now know that these assumptions were tied to patterns of racial bias and entrenched ideas about who was the wrong kind of addict. But the problem is not simply that the law has been historically unfair—though, of course, it has. The problem is that the massive disparities in law enforcement persist.[36] In a sense, this is not surprising. Unjust biases shape every aspect of American life, ranging from education to healthcare to incarceration rates. These problems must be addressed if our aim is harm reduction and, eventually, the pursuit of individual and social healing.

In some ways, opioid addiction seems like a great leveler. Anyone can become addicted, and overdoses can happen to anyone. So, we should be able to focus policies on harm reduction in ways that operate without recourse to racial or class-based biases. As a society, though, we have never been good at crafting solutions to social problems that take the whole community into account. Even attempts at fairness are often fragile. This issue is especially pronounced in drug enforcement strategies and differences in access to medical resources, which continue to be shaped by racial and class biases. Put bluntly, how you are likely to be treated after an overdose critically depends on where you live and the color of your skin. This is a deep problem that needs to be addressed. In doing so, fairness should not be assumed as a default. It should be seen as something that must be achieved. This will require acknowledging differences in biological vulnerabilities toward opioid abuse, but it will also require attending to the social and cultural factors that shape the likelihood of addiction and impact the availability and quality of care in the context of addiction.

Increasing knowledge of the opioid epidemic, alongside advances in preventing or reversing overdoses, has started to reshape attitudes toward addiction. People appear to be less likely to treat addiction as a moral vice, and there seems to be a growing emphasis on harm reduction.[37] But what does it mean to make our primary goals minimizing harm and helping people pursue the resources they need for rehabilitation or habilitation?

CONCLUSION: REMEMBER, WE'RE ALL IN THIS TOGETHER

Recovery is often possible. But it is never achieved all at once, and it rarely occurs in a linear way. There are many routes to recovery. People who pursue the same path might not achieve the same milestones at the same time. This is what we should expect since sobriety is a habit that must be cultivated. This can be a life-long process that requires a great deal of aid and support, but importantly, the task of cultivating this habit often gets easier over time. Eventually, feelings of joy and relief from pain will arise without drugs as the brain begins to release endogenous opioids in a wider range of contexts.

The path to recovery often begins with attempts to imagine a different way of living. This might occur after an overdose makes it seem necessary to pursue another life. It might occur because someone realizes they have abandoned their friends or family. Or it might occur when they witness someone else's pain and loss. In each case, the attempt to imagine a better life is only a first step. After all, addiction is a deep form of physiological dysregulation, which bends thought and behavior in a specific direction, and attempts to imagine better ways of living are not immune to these effects. The sadness, loss, and tragedy of addiction have many different avenues of expression, but they often lead people who are addicted to opioids to *feel vulnerable* or feel like they are compelled to pursue worse options even though better ones are available. As we argued in an earlier chapter, the continued experience of temptation can lead to overactivation of neural resources. This makes it hard to consider alternatives or look beyond the most salient feelings of pain and distress. Against this backdrop, there are three things we should keep in mind in exploring a path out of our current opioid epidemic.

1. The pain of withdrawal is real, as are the pains of loss, isolation, and grief that often accompany addiction. This is where harm reduction strategies have focused their attention. There is nearly universal consensus that naloxone should be broadly available to manage overdoses.[38] There is also growing support for "some form of legal supply for confirmed opiate addicts."[39] If the dispersal of opioids to people who are addicted is carried out by healthcare providers, this might reduce overdose risk by preventing people from buying heroin that is cut with fentanyl, for example. This need not take the form of the

continued supply of painkillers from a physician. The aim should be making sure that people who are addicted are not pursuing worse options because they are the only ones available. To be clear, currently available opioids are all complicated and dangerous substances, so this will not be a panacea. But given the scale of the challenges we face, it seems like one part of a plausible path forward.

2. Harm reduction has become a key pillar of public health. In recent years, a few attempts have been made to pursue decriminalization as a harm-reduction strategy. As Maia Szalavitz argues, decriminalization seemed like a long shot in the 1990s.[40] In our current context, a broad attempt at decriminalization might provide a necessary supplement to existing attempts to manage addiction. Policing and incarceration rates show deep effects of racial and class biases.[41] Moreover, as Szalavitz rightly notes: "Jail doesn't treat addiction successfully; within jail, treatment is rarely available; incarceration increases overdose death risk; arrest is not a good way to determine who most needs care; and criminalization wastes enormous amounts of money that might otherwise fund effective help."[42] That said, the pursuit of decriminalization must be pursued in parallel with other forms of harm reduction, and this can only be one part of the broader project of pursuing a better world in the face of the current opioid epidemic.

3. Health plays a crucial role in the willingness to pursue our hopes and perceive our lives as meaningful. This doesn't mean there is only one way to be healthy. In fact, there are many ways to be embodied, so there are many ways to be healthy.[43] Contact and social reassurance are essential to health for most people, and indeed, most primates.[44] In part, this is because contact driven by care often facilitates the release of oxytocin and prolactin. While the story is complex, it seems that this

can often dampen distress and help us to perceive others as a source of social comfort.[45] But in part, this is because diverse evolutionary and developmental processes support tendencies toward collaboration in our species and put us in a position to pursue mutual aid and mutual support. Of course, the relations that support contact and social assurance differ across cultures and individuals, but in general, our brains seem to work best in the company of those we love.[46] So, while pursuing health is important, we need a conception of health that is focused on "embodied care" and the cultivation of social connections.[47]

Drawing these threads together, a perspective comes into view that highlights the dense connection between the experience of addiction and the structure of our social world. Some contexts make it easier to make bad decisions; others make it easier to make good choices. Therapeutic interventions should take advantage of this fact, attempt to flip the paradigm and construct better ways for people to live and act together. This is a tall order! But research into strategies for promoting and solidifying recovery is beginning to take place in communities in ways that highlight the significant role of recovery networks.[48] Such networks might take the form of programs like Narcotics Anonymous. However, as we noted earlier, sitting together, talking with one another, and committing to one another can have a profound effect on experience. Commitments to building habits and relations of trust where there seem to be few alternatives might help as well.

Note, however, that from this perspective, recovery requires more than treatment for an addicted body. Put much too simply, recovery is never *just* a matter of getting the relevant chemicals out of the system. It requires learning to rely upon others (perhaps for the first time) and learning about the promises and the burdens of life within a community that can help pull one out of the whirlpool of addiction.[49] Finally, it requires helping people cultivate capacities for imagining better ways to live and helping them develop the habits that will bring this way of living into practice. This isn't just a medical intervention—it's an intervention that acknowledges the complex connections between physical, mental, and social health. This is what is necessary to help people take hold of their lives and to help them construct richer possibilities for more autonomous action.

Autonomy might seem like an impossible dream for someone who is addicted to opioids. However, we must remember that no one ever pursues autonomy on their own.[50] We are a social species—we get by with the aid and support of others, we lean on each other's strengths, we learn from one another, we seek out help from those we trust, and when things go badly, we look to others for help. It is often assumed that people make worse choices in groups, and the rise of social media, along with the prevalence of disinformation, seems to support this claim. A commitment to deciding together can help us to make better decisions when the conditions are right.[51] In groups, especially small groups that foster trust and shared commitments, we can explore problems and potential solutions from multiple perspectives. We can also discover problems with our positions as we attempt to bring others to our side. Of course, this is not easy, and it requires shared commitments to making sure every voice is heard and every piece of information is considered.[52] When we attempt to build a life together, such commitments can arise in a fairly seamless way. This is especially important in the context of recovery, where imagining a better future or a better way of living can be almost impossible without help.

This doesn't mean the whole story is social, nor does it mean that all of the work that must be done is a matter of rebuilding a social world. But the path toward recovery will often involve cultivating a richer awareness of what can be shared with others and an openness to the possibilities that others can point toward. It is hard to know where to start in cultivating these attitudes. After all, anger and hopelessness are common in our world, as are experiences of isolation and social fragmentation.[53] Without support from others and a shared sense of purpose, the ability to imagine relief will often evaporate. By contrast, stable patterns of social engagement and higher degrees of social integration can sometimes help us avoid the drift toward more negative spaces, and trusted communities can provide a buffer against relapses and a source of support for getting back on the path to recovery when relapses occur.

This shouldn't be surprising. Most of us have sought the assistance of friends or family after we started projects and stalled out because it was unclear how to go on or succeed. Moreover, most of us know that communities can support better ways of living and acting by helping us to stay motivated. You might find your community in a church, a Buddhist sangha, a group that plays basketball together, or a group that discusses movies and music—or you might find it in a support group or a twelve-step program.[54] Such communities can provide a valuable source of aid and support. This kind of support is crucial in the context of addiction and the pursuit of rehabilitation or habilitation. Recall that addiction both supports and is shaped by depression—and depression both supports and feeds upon isolation.[55] As Emile Durkheim noted long ago, suicide risk increases when a person feels isolated and when they are not integrated into their community.[56] We need not accept his account of this effect, but he was right to claim that social integration and trust are often necessary for people to seek help, and he was right that the community can ameliorate social pain. Of course, communities can also exacerbate social pain—so it's important to build our social lives around robust networks of mutual aid and mutual support. Our chances of surviving and flourishing improve the more we connect with others in supportive and meaningful ways.[57]

Cultivating social hope is hard work. It requires believing things can be better and the support of a community that believes things can change for the better. Cultivating social hope is even harder for someone who is dealing with addiction, and it can be incredibly hard for their friends and loved ones, who may have been hurt in profound ways, and who (rightly) find it hard to believe that things can be different. But if there is only one thing that we can be sure of, it's that change is a pervasive feature of our world. Learning is a pervasive feature of human cognition, and epigenetic changes occur in all of the neural tissues that are important for recovery.[58] That said, such changes always depend on the connection to the larger social and ecological world. This matters because social hope is a tool for guiding such change in a better direction.

CONCLUSION

Pursuing Freedom

Some people pursue opioids to alleviate physical or psychological pain. Some people seek oblivion, and some chase an opium "high," instead of trying to avoid aversive experiences. Most people who use opioids will never become addicted, and many researchers observe that the risk of addiction decreases as people reach their thirties. Some people seem to have an easier time with recovery, though it's not clear why; others who pass through withdrawal—often with a great deal of help—regain control of their lives and never look back. Relapsing is also common, but the patterns of relapsing are not uniform. Finally, there are far too many people who die, regardless of how they first became involved with opioids.

It is unlikely that any single factor will explain all of the ways that addiction unfolds. As we have argued throughout this book, addiction is a form of cephalic dysregulation, which is shaped by social regularities that shape patterns of thought and action. It typically yields a narrowing of focus, and it often *seems* to support compulsions that overpower other goals.[1] As we argued in an earlier chapter, this does not mean that cognition is hijacked by addiction, but in this conclusion, we explore the implications of our perspective on opioid addiction for pursuing freedom and self-control.

Addiction often *feels like* a loss of choice. Moreover, luck and genetics appear to shape the likelihood of addiction, as do histories of trauma and stress and cultural situations that foster oppression and exclusion. None of these is the sole or even primary reason for addiction. It is clear that thought and action are always constrained by our biology, our learning histories, and our social and material environments.[2] Interactions among biology, embodiment, and social environments can guide people toward specific choices and away from others. Addiction seems to impose a further constraint on choices by promoting patterns of thought and action that make bad choices more likely. Intellectuals like Daniel Wegner and Robert Sapolsky suggest that such phenomena support the hypothesis that freedom is illusory.[3] By contrast, someone like Daniel Dennett suggests that such issues should be reframed to acknowledge the expanded capacities for choice that have evolved within typical members of our species.[4] Our perspective offers a middle path between these claims.

We doubt that it is useful to treat people as automatons or as beings who act on the basis of perfect freedom.[5] The ability to choose is a feature of our biology. Like all biological phenomena, it is interpreted and experienced through a cognitive and cultural lens. The world always affords *some* opportunities for choice, though, even when it is difficult to know how our choices are constrained and which choices lie beyond our control. As Richard Holton and Kent Berridge rightly note, even "the intensity and power of an addictive desire does not mean that individuals are automata, powerless spectators moved by their desires."[6] But this is only part of the story. Given the complexity of the causes and conditions that produce thought and behavior, we must all live with uncertainty about how much control we actually have. Choice might often seem to imply the freedom to travel along different paths, but each choice is shaped by a range of historical and situational constraints. Some choices feel more free, others feel more forced, and sometimes it feels like there aren't really any choices to make. Experiences of having fewer choices or more highly constrained options might be common in the context of addiction, but there is never a clear distinction to be drawn between choices that are free and ones that are massively constrained. This is the human predicament.

This might seem to lead us to a perspective that is not dissimilar from the free will skeptics that we mentioned above, but the world is far too complex for a constructed binary like free versus constrained to apply. We all make bad choices in moments when we are distracted or overwhelmed, and *cravings* sometimes lead us to make judgments that are less consistent with our long-term values.[7] For example, consider the kinds of purchases you make when you go shopping hungry. Hunger can alter our decisions, at least to some extent. This is a familiar fact about human experience. Most of the time, however, people can rely upon cognitive heuristics and problem-solving strategies that are good enough to navigate the challenges and opportunities they typically face.[8] Of course, the impact of opioid cravings is far more pronounced.[9] In this context, excessive wanting can radically change behavior,[10] and behavioral control can be compromised or dramatically reduced.[11] After all, addiction shapes how people seek pleasure, avoid pain, and behave—more or less—rationally in response to various challenges and opportunities. Our point is simply that this is a situation we can all understand, even if we only find faint echoes of such experiences in ourselves.[12]

As we suggested in an earlier chapter, addiction yields a form of *akrasia*. In classical Greek traditions, *akrasia* was often described as a conflict between reason and desire where emotion or desire leads someone to pursue a worse option, though a better path is available. Similar phenomena were discussed in South

Asian philosophical traditions, where people are often described as acting from ignorance, against their better nature, or against their best judgment. In each of these contexts, we find someone who knows what is right, but fails to do it; knows what is wrong, but fails to stop; or acts in ways they regret. In both philosophical traditions, these are presented as patterns of choice and will that have been bent toward problematic pursuits.

Intriguingly, philosophical approaches that have centered on these phenomena have often proposed strategies for cultivating freedom, which highlight the importance of habituation. Sometimes, this means learning to avoid choices that will be pleasurable in the short run to ensure flourishing in the long run. Sometimes, it requires recognizing that decision making is continuous with effort and building more resilient capacities to stay the course when the going gets tough. The key thing to remember is that we all make worse decisions in some contexts, so cultivating freedom always requires organizing our habits and environments to play to our strengths and minimize our vulnerabilities.

Consider a proposal that was defended by the seventeenth-century Dutch philosopher Benedict de Spinoza. He held that people mistakenly believe they are free because they are aware of their desires and motivations but ignorant of the causes that lead them to desire and pursue the things they do.[13] It is easy to avoid asking difficult questions about what causes particular behaviors and what it takes to change them. Spinoza held that we often rely upon habits of thought, which lead us to pursue worse options when there are better ones available.[14] He was a hopeful philosopher though, and he suggested that cultivating a better understanding of the causes of our motivations can help us to change our strategies for exploring the world while also minimizing the effects of harmful motivations. Specifically, he claimed that we can figure out which interactions increase our capacities to act well and we can set up the conditions—both internally and externally—that will allow us to pursue individual and collective liberation.

A slightly different perspective was proposed by South Asian Buddhist traditions, which focus on eliminating suffering and disquietude. These perspectives acknowledge that our understandings of the world and the actions that flow from such understandings are shaped by complex networks of causes and conditions. They argued that although we have some control over our actions, we can gain more control by cultivating better habits and practical dispositions.[15] The core assumption is that it is possible to reshape our tendencies to grasp at the self as a locus of autonomous action and to ignore our deeply pervasive forms of sociality. This requires practices of mental cultivation that could entrench new habits of thought and action and guide us toward individual and collective liberation.[16]

We might not want to follow Spinoza or these Buddhist philosophers, but the parallel between these two frameworks points the way toward an important insight. To cultivate social hope, it is often necessary to search for buried truths, come to terms with ugly and complex realities, and express those realities to others. Sometimes, this means adopting a more critical perspective on oneself. Sometimes, it means acknowledging that you are the cause of harm to others. Like all exercises in truth seeking, this one can be quite painful. In many cases, it can be hard for someone who is addicted to tell which things they have done are right and which are wrong—this is one of the pernicious effects of having thought and attention bent toward the pursuit of opioids. For someone who is addicted, it might be difficult to know which desires to treat as their own and which to understand as reflecting drug-induced cravings. It might also be hard to know which memories are factual and which reflect distorted and bent memories that were shaped by the drugs they pursued. This will make it difficult to know what the truth is—and this is a critical point where people need support from others to find their way back to a better way of being in the world.

For someone who is addicted, it will often be hard to avoid situations, people, and stressors that lead to opioid use. Moreover, it will often be difficult to avoid the peer pressure and isolation that support relapses. Thus, recovery often requires constructing new communities or reconstructing collapsed ones. This is incredibly difficult, but it is often the only way to reconnect with the sources of encouragement that can help when failure seems inevitable. Addiction has many forms, but opioids make it clear that we are all in this together. It is in all of our interests to respond to the needs of others and recognize them as *humans* whose lives and well-being *should be* matters of concern to me, you, and us. There is no guarantee that the sense of being in this together will help us create the necessary communities, but it might help us all see our roles in helping the people in our communities to thrive and flourish where possible. But we need to go one step further; we must attempt to understand all human lives as rich sources of possibility. By cultivating a reasonable degree of awe toward the diversity and beauty of people, we can begin to reveal pathways for creatively transforming experiences.[17]

At this point, the necessity of action and cultivating new habits come more clearly into view. As we argued in an earlier chapter, the habits we build help us to focus on what matters. But the cultivation of habits can be pursued intentionally or shaped through repeated activities that reflect less intentional attempts to manage the challenges and opportunities we encounter. Cultivating habits that sustain "self-control" often requires embedding ourselves within environments that reward "self-control." In the context of addiction, this might take the form of checking into a rehab facility to prevent buying more drugs, building a connection with a friend who can be called or texted for support, or building supportive therapeutic communities that can offer reminders of commitments to pursue recovery. These strategies aim to bring future benefits to mind where cravings or the pain of withdrawal make it easy to lose sight of a long-term goal. These choices constrain future behaviors, but in doing so, they also create a kind of freedom.

Sometimes, we can set our future selves up for success by using various "self-binding" strategies. For example, precommitting to an action can decrease the likelihood of pursuing an immediate reward.[18] By specifying a precise, situation-specific action plan, we can change the stakes of acting and change "the task of self-control we confront."[19] This can cause our goals and plans to have a more significant impact on present motivation.[20] One method for doing this is by rehearsing simple if-then plans that specify precise trigger cues and behavioral responses—this appears to enact a basic form of action control, which can shield behavior from temptation and distraction and counteract the effects of habitual patterns of thought and behavior.[21] This only works where we already have strong commitments to particular goals, but it can help us to maintain control over behavior by linking cues that we consider critical to an action to a behavioral response that we already believe to be relevant to achieving a goal.[22]

To the extent that we can anticipate future challenges, we can find ways to limit the effect of the harmful impulses that are likely to arise in the future. Such practices are especially important for someone who is just beginning the process of recovery. After all, it takes time to get used to any new way of life, and kicking an addiction involves changing nearly everything about the way one expects to relate to the world. By finding ways to bring future situations to mind and remembering the self-binding actions that have been settled on, it is possible to use such commitments to preserve a kind of continuity through change.

This brings us to a final and important point. The chaos of bouncing in and out of rehab, onto and off the street, overshadows one final, strange feature of addiction. In the context of addiction, life is often dangerous, but often, it can be profoundly boring. The all-consuming cycle of craving, pain, and inadequate relief from pain makes it more and more difficult to *want* to go on. Perhaps it is unsurprising, then, that theater and games—which promote expression and creativity—have proven useful in therapeutic contexts.[23] This is one way of trying to embody another life and seeing how things could be different. Whatever the path might be, it is important to find ways of cultivating positive emotional responses, as spirals of pain and depression make it hard to keep going. This will often support relapses. While this might not work for everyone, music, dancing, and social laughter seem to be particularly fruitful triggers of endogenous opioids that might guard against these patterns of relapse.[24]

There are numerous other issues that could be addressed at this point. But for now, the key thing that we want to highlight is that our choices often reflect the structure of our social and material environment. This is a fact about our biology, psychology, and sociality. The cultural contexts where we find ourselves are an amalgamation of the habits of many people.[25] These contexts shape our expectations, and they impact our actions and our ongoing pursuits.[26] Finding new ways to live and act will often require creating new patterns of behavior that can fit into those that have already been established by others. Fortunately, we are rabid social niche constructors and socially shaped shapers of our environment.[27] We can accomplish great changes in ourselves, our bodies, and our social worlds. Doing so requires time, and it often requires creating the right context for producing the changes we want to see. We express our goals to ourselves and others, and the way we do so shapes our behavior and impacts the way that we construct the self and the world that we want to inhabit.[28] This is the kind of choice that is essential to how we understand ourselves.

With all of these claims in mind, let us close by reiterating the fact that the range of physiological, psychological, and phenomenological factors that must be considered in thinking about opioid addiction is simply astounding. We know that the vulnerability to opioid addiction has been present across times and cultures, and we know that people from diverse social classes, with a wide range of social identities, have been affected.[29] Just as importantly, we know that numerous social, political, and cultural factors shape our dominant practices for making sense of addiction. For example, who counts as *using* opioids and who counts as *abusing* them changes often. Some people are perceived as experiencing an unfortunate "state of physical adaptation in which the drug is required for normal functioning."[30] Others with similar experiences and different social identities are treated as "addicts." This term carries a lot of psychological and phenomenological baggage, and when it is used, it is often intended to signal a moral failing that someone brings upon themself.[31] In these contexts and many others, it is necessary to examine the cultural constraints that shape our understanding of addiction and the constraints that lead people to engage with opioids.

Thankfully, prefrontal lobotomies are no longer employed to manage morphine addiction.[32] We now recognize that prefrontal networks play a crucial role in our understanding of the world and our reactions to it. It is often claimed that the prefrontal cortex is the seat of self-control and the ability to orient toward our goals. Likewise, it is often suggested that the prefrontal cortex

is crucial to managing addiction because of the role this structure plays in inhibiting those habits that pull us toward worse
options.[33] At most, however, this can only be part of a plausible story. Every change that occurs in a brain occurs in a larger
cultural context.[34] This is one of the reasons why we should not
place more weight on biological approaches to addiction than
current data can support.[35] We should also remain circumspect
about attempts to treat addiction solely through chemical interventions.[36] Medical interventions will often help, and access to
medication plays a crucial role in harm reduction. However, we
should avoid overmedicalizing the human condition, remember that recovery often involves changing relationships between
a person and their environment, and all take time to reflect on
what we can do—as individuals and as a society—to ameliorate
the inequity, unfairness, and disrespect that pervade the current
opioid epidemic.

NOTES

INTRODUCTION: A COMPLEX STORY ABOUT BIOLOGY AND CULTURE

1. Scott Atran and Douglas Medin, *The Native Mind and the Cultural Construction of Nature* (MIT Press, 2008).
2. Don Ross, "Addiction Is Socially Engineered Exploitation of Natural Biological Vulnerability," in *Evaluating the Brain Disease Model of Addiction*, ed. Nick Heather et al. (Routledge, 2022), 359–72.
3. Charles Darwin, *The Origin of Species* (Mentor Book, 1859/1972).
4. Charles Darwin, *Incectivorous Plants* (John Murray, 1875).
5. Li Guo et al., "The Opium Poppy Genome and Morphinan Production," *Science* 362, no. 6412 (2018): 343–47; Yiheng Hu et al., "The Genome of Opium Poppy Reveals Evolutionary History of Morphinan Pathway," *Genomics, Proteomics and Bioinformatics* 16, no. 6 (2018): 460–62; Dan Larhammar et al., "Early Duplications of Opioid Receptor and Peptide Genes in Vertebrate Evolution," *Annals of the New York Academy of Sciences* 1163, no. 1 (2009): 451–53.
6. Jacobus C. De Roode, Thierry Lefèvre, and Mark D. Hunter, "Self-Medication in Animals," *Science* 340, no. 6129 (2013): 150–51; Daniel H. Janzen, "Why Fruits Rot, Seeds Mold, and Meat Spoils," *American Naturalist* 111, no. 980 (1977): 691–713; Sabrina Krief et al., "Bioactive Properties of Plant Species Ingested by Chimpanzees (Pan troglodytes schweinfurthii) in the Kibale National Park, Uganda," *American Journal of Primatology* 68, no. 1 (2006): 51–71.

7. Michael A. Huffman, "Animal Self-Medication and Ethno-Medicine: Exploration and Exploitation of the Medicinal Properties of Plants," *Proceedings of the Nutrition Society* 62, no. 2 (2003): 371–81.

8. H. C. Morrogh-Bernard et al., "Self-Medication by Orang-Utans (Pongo pygmaeus) Using Bioactive Properties of Dracaena Cantleyi," *Scientific Reports* 7, no. 1 (2017): 16653.

9. Sarah Blaffer Hrdy, *Mothers and Others: The Evolutionary Origins of Mutual Understanding* (Harvard University Press, 2009); David Premack and Ann James Premack, "Origins of Human Social Competence," in *The Cognitive Neurosciences*, ed. Michael S. Gazzaniga (MIT Press, 1995), 205–18.

10. Hazel Rose Markus and Shinobu Kitayama, "Cultures and Selves: A Cycle of Mutual Constitution," *Perspectives on Psychological Science* 5, no. 4 (2010): 420–30; Qi Wang, *The Autobiographical Self in Time and Culture* (Oxford University Press, 2013).

11. Rebecca Wragg Sykes, *Kindred: Neanderthal Life, Love, Death and Art* (Bloomsbury, 2020).

12. Peter J. Richerson and Robert Boyd, "Rethinking Paleoanthropology: A World Queerer Than We Supposed," *Evolution of Mind, Brain, and Culture* (2013): 263–302.

13. Alison Gopnik, "Childhood as a Solution to Explore–Exploit Tensions," *Philosophical Transactions of the Royal Society B* 375, no. 1803 (2020): 20190502.

14. Denise Dellarosa Cummins and Robert Cummins, "Biological Preparedness and Evolutionary Explanation," *Cognition* 73, no. 3 (1999): B37–B53; Paul Rozin and James W. Kalat, "Specific Hungers and Poison Avoidance as Adaptive Specializations of Learning," *Psychological Review* 78, no. 6 (1971): 459.

15. Stephen Mithen, *The Prehistory of the Mind: The Cognitive Origins of Art and Science* (Thames and Hudson, 1996); Peter J. Richerson and Robert Boyd, *Not by Genes Alone: How Culture Transformed Human Evolution* (University of Chicago Press, 2008).

16. Herbert A. Simon, "The Architecture of Complexity," *Proceedings of the American Philosophical Society* 106, no. 6 (1962): 470–73; Herbert A. Simon, *Models of Bounded Rationality: Economic Analysis and Public Policy* (MIT Press, 1982); Gerd Gigerenzer, *Adaptive Thinking* (Oxford University Press, 2000).

17. Bryce Huebner, "Planning and Prefigurative Practice," in *The Philosophy of Daniel Dennett* (Oxford University Press, 2017), 295–327.

18. Paul Rozin, "The Socio-Cultural Context of Eating and Food Choice," in *Food Choice, Acceptance and Consumption*, ed. H. L. Meiselman and H. J. H. MacFie (Springer, 1996), 83–104.

19. Robin Dunbar, *Human Evolution: Our Brains and Behavior* (Oxford University Press, 2016); Robin Dunbar et al., "Social Laughter Is Correlated with an Elevated Pain Threshold," *Proceedings of the Royal Society B: Biological Sciences* 279, no. 1731 (2012): 1161–67; John Sabini and Maury Silver, *Moralities of Everyday Life* (Oxford University Press, 1982).

20. Zoe R. Donaldson and Larry J. Young, "Oxytocin, Vasopressin, and the Neurogenetics of Sociality," *Science* 322, no. 5903 (2008): 900–904.

21. Sandra Manninen et al., "Social Laughter Triggers Endogenous Opioid Release in Humans," *Journal of Neuroscience* 37, no. 25 (2017): 6125–31; Lauri Nummenmaa et al., "Social Touch Modulates Endogenous μ-Opioid System Activity in Humans," *NeuroImage* 138 (2016): 242–47.

22. Bryce Huebner and Jay Schulkin, *Biological Cognition* (Cambridge University Press, 2022).

23. Joshua B. Tenenbaum et al., "How to Grow a Mind: Statistics, Structure, and Abstraction," *Science* 331, no. 6022 (2011): 1279–85.

24. Charles Sanders Peirce, "The Fixation of Belief," *Popular Science Monthly*, no. 12 (1877): 1–15; Charles Sanders Peirce, *Reasoning and the Logic of Things*, ed. Kenneth Lane Ketner and Hillary Putnam (Harvard University Press, 1899/1992).

25. Paul Mellars, "Why Did Modern Human Populations Disperse from Africa ca. 60,000 Years Ago? A New Model," *Proceedings of the National Academy of Sciences* 103, no. 25 (2006): 9381–86.

26. Christopher Boehm, *Moral Origins: The Evolution of Virtue, Altruism, and Shame* (Soft Skull, 2012); Kim Sterelny, *The Pleistocene Social Contract: Culture and Cooperation in Human Evolution* (Oxford University Press, 2021).

27. Peter Kropotkin, *Mutual Aid: A Factor of Evolution* (Black Rose, 1904/2021).

28. Dunbar, *Human Evolution*.

29. Stephen Jay Gould, *The Structure of Evolutionary Theory* (Cambridge University Press, 2002).

30. Kropotkin, *Mutual Aid*.

31. Ross, "Addiction Is Socially Engineered Exploitation."

32. Ross, "Addiction Is Socially Engineered Exploitation."

33. Michael L. Power and Jay Schulkin, *The Evolution of Obesity* (Johns Hopkins University Press, 2009).

34. Peter Sterling, *What Is Health? Allostasis and the Evolution of Human Design* (MIT Press, 2020).

35. Jay Schulkin and Peter Sterling, "Allostasis: A Brain-Centered, Predictive Mode of Physiological Regulation," *Trends in Neurosciences* 42, no. 10 (2019): 740–52.

36. Andy Clark, *Surfing Uncertainty: Prediction, Action, and the Embodied Mind* (Oxford University Press, 2015); Philip R. Corlett, Aprajita Mohanty, and Angus W. MacDonald III, "What We Think About When We Think About Predictive Processing," *Journal of Abnormal Psychology* 129, no. 6 (2020): 529.

37. Giovanni Pezzulo, Thomas Parr, and Karl Friston, "The Evolution of Brain Architectures for Predictive Coding and Active Inference," *Philosophical Transactions of the Royal Society B* 377, no. 1844 (2022): 20200531.

38. Sterling, *What Is Health?*; Peter Sterling and Simon Laughlin, *Principles of Neural Design* (MIT Press, 2015).

39. Huebner and Schulkin, *Biological Cognition*.

40. Peirce, "The Fixation of Belief"; Peirce, *Reasoning and the Logic of Things*.

41. Charles P. O'Brien et al., "Penn/VA Center for Studies of Addiction," *Neuropharmacology* 56 (2009): 44–47.

42. Ann M. Graybiel, "The Basal Ganglia and Chunking of Action Repertoires," *Neurobiology of Learning and Memory* 70, nos. 1–2 (1998): 119–36.

43. Kent C. Berridge and Terry E. Robinson, "What Is the Role of Dopamine in Reward: Hedonic Impact, Reward Learning, or Incentive Salience?," *Brain Research Reviews* 28, no. 3 (1998): 309–69; Kent C. Berridge, "Is Addiction a Brain Disease? The Incentive-Sensitization View," in *Evaluating the Brain Disease Model of Addiction* (Routledge, 2022), 74–86.

44. Natasha Dow Schüll, "Addiction by Design: Machine Gambling in Las Vegas," in *Addiction by Design* (Princeton University Press, 2012).

45. Zoey Lavallee, "Affective Scaffolding in Addiction," *Inquiry* (2023): 1–29.

46. Sterling. *What Is Health?*

47. David T. Courtwright, *Forces of Habit: Drugs and the Making of the Modern World* (Harvard University Press, 2001).

48. *contra* Markus Heilig et al., "Addiction as a Brain Disease Revised: Why It Still Matters, and the Need for Consilience," *Neuropsychopharmacology* 46, no. 10 (2021): 1715–23; Alan I. Leshner, "Addiction Is a Brain Disease, and It Matters," *Science* 278, no. 5335 (1997): 45–47; Nora D. Volkow, George F. Koob, and A. Thomas McLellan, "Neurobiologic Advances from the Brain Disease Model of Addiction," *New England Journal of Medicine* 374, no. 4 (2016): 363–71.

49. Julie Netherland and Helena Hansen, "White Opioids: Pharmaceutical Race and the War on Drugs That Wasn't," *BioSocieties* 12, no. 2 (2017): 217–38.

50. David T. Courtwright, *Dark Paradise: A History of Opiate Addiction in America* (Harvard University Press, 1982); Courtwright, *Forces of Habit*; Nancy D. Campbell, *OD: Naloxone and the Politics of Overdose* (MIT Press, 2020); David Herzberg, *White Market Drugs: Big Pharma and the Hidden History of Addiction in America* (University of Chicago Press, 2020).

51. Gopnik, "Childhood as a Solution to Explore–Exploit Tensions."

52. Mary Jeanne Kreek, Brian Reed, and Eduardo Butelman, "Current Status of Opioid Addiction Treatment and Related Preclinical Research," *Science Advances* 5, no. 10 (2019): eaax9140.

53. Volkow, Koob, and McLellan, "Neurobiologic Advances from the Brain Disease Model of Addiction"; George F. Koob and Jay Schulkin, "Addiction and Stress: An Allostatic View," *Neuroscience & Biobehavioral Reviews* 106 (2019): 245–62.

54. Maia Szalavitz, *Unbroken Brain: A Revolutionary New Way of Understanding Addiction* (St. Martin's, 2016).

55. Theodore J. Cicero et al., "The Changing Face of Heroin Use in the United States: A Retrospective Analysis of the Past 50 Years," *JAMA Psychiatry* 71, no. 7 (2014): 821–26.

56. Edgar Allen Poe, as quoted in Szalavitz, *Unbroken Brain*, 20

57. Herzberg, *White Market Drugs*; Jonathan S. Jones, "Opium Slavery," *Journal of the Civil War Era* 10, no. 2 (2020): 185–212.

58. Patrick Radden Keefe, *Empire of Pain: The Secret History of the Sackler Dynasty* (Anchor, 2021).

59. Campbell, *OD*; Herzberg, *White Market Drugs*.

60. George F. Koob and Michel Le Moal, *Neurobiology of Addiction* (Elsevier, 2006).

61. Koob and Le Moal, *Neurobiology of Addiction*.

62. Michael Kuhar, *The Addicted Brain: Why We Abuse Drugs, Alcohol, and Nicotine* (FT Press, 2011).

63. Koob and Le Moal, *Neurobiology of Addiction*; Koob and Schulkin, "Addiction and Stress."

64. Bruce S. McEwen, "Allostasis and the Epigenetics of Brain and Body Health over the Life Course: The Brain on Stress," *JAMA Psychiatry* 74, no. 6 (2017): 551–52

65. Campbell, *OD*; Herzberg, *White Market Drugs*.

1. EFFORT AND DECISION MAKING

1. Substance Abuse and Mental Health Services Administration, "Key Substance Use and Mental Health Indicators in the United States: Results from the 2021 National Survey on Drug Use and Health," HHS Publication No. PEP22-07-01-005, NSDUH Series H-57 (Center for Behavioral Health Statistics and Quality, Substance Abuse and Mental Health Services Administration, 2022), https://www.samhsa.gov/data/report/2021-nsduh-annual-national-report.

2. Travis N. Rieder, *In Pain: A Bioethicist's Personal Struggle with Opioids* (Harper Collins, 2019).

3. Carl L. Hart, *Drug Use for Grown-Ups: Chasing Liberty in the Land of Fear* (Penguin, 2021).

4. David Herzberg, *White Market Drugs: Big Pharma and the Hidden History of Addiction in America* (University of Chicago Press, 2020).

5. Virginia Berridge and Griffith Edwards, *Opium and the People: Opiate Use in Nineteenth-Century England* (Free Association Books, 1981).

6. David T. Courtwright, *Dark Paradise: A History of Opiate Addiction in America* (Harvard University Press, 1982); David T. Courtwright, *Forces of Habit: Drugs and the Making of the Modern World* (Harvard University Press, 2001); David T. Courtwright, *The Age of Addiction: How Bad Habits Became Big Business* (Harvard University Press, 2019).

7. Barry J. Everitt et al., "Neural Mechanisms Underlying the Vulnerability to Develop Compulsive Drug-Seeking Habits and Addiction,"

Philosophical Transactions of the Royal Society B: Biological Sciences 363, no. 1507 (2008): 3125–35.

8. Barry J. Everitt and Trevor W. Robbins, "Drug Addiction: Updating Actions to Habits to Compulsions Ten Years On," *Annual Review of Psychology* 67, no. 1 (2016): 23–50.

9. William James, *The Principles of Psychology*, vol. 1 (Dover, 1890/1952), 104ff.

10. Kyle S. Smith and Ann M. Graybiel, "Habit Formation," *Dialogues in Clinical Neuroscience* 18, no. 1 (2016): 33–43; and Andreas K. Engel, Karl J. Friston, and Danica Kragic, eds., *The Pragmatic Turn: Toward Action-Oriented Views in Cognitive Science*, vol. 18 (MIT Press, 2016).

11. John Dewey, "The Reflex Arc Concept in Psychology," *Psychological Review* 3, no. 4 (1896): 357; Wendy Wood and David T. Neal, "A New Look at Habits and the Habit-Goal Interface," *Psychological Review* 114, no. 4 (2007): 843; Engel, Friston, and Kragic, *The Pragmatic Turn*.

12. Kent C. Berridge, "Affective Valence in the Brain: Modules or Modes?," *Nature Reviews Neuroscience* 20, no. 4 (2019): 225–34.

13. Wood and Neal, "A New Look at Habits and the Habit-Goal Interface."

14. Kent C. Berridge and Terry E. Robinson, "What Is the Role of Dopamine in Reward: Hedonic Impact, Reward Learning, or Incentive Salience?," *Brain Research Reviews* 28, no. 3 (1998): 309–69; Charles R. Gallistel, *The Organization of Action* (Earlbaum, 1980); Jay Schulkin, *Effort: A Behavioral Neuroscience Perspective on the Will* (Erlbaum, 2007).

15. Anthony Dickinson and Bernard Balleine, "Motivational Control of Goal-Directed Action," *Animal Learning & Behavior* 22, no. 1 (1994): 1–18; James, *The Principles of Psychology*; Russell Poldrack, *Hard to Break: Why Our Brains Make Habits Stick* (Princeton University Press, 2021).

16. Engel, Friston, and Kragic, *The Pragmatic Turn*.

17. James, *The Principles of Psychology*, vol. 2, 252.

18. Antoine Bechara, "Decision Making, Impulse Control and Loss of Willpower to Resist Drugs: A Neurocognitive Perspective," *Nature Neuroscience* 8, no. 11 (2005): 1458–63.

19. Schulkin, *Effort*.

20. Jon Elster, *Ulysses Unbound* (Cambridge University Press, 2000); Brendan Dill and Richard Holton, "The Addict in Us All," *Frontiers in Psychiatry* 5 (2014): 139.

21. James, *The Principles of Psychology*.

22. Bryce Huebner and Jay Schulkin, *Biological Cognition* (Cambridge University Press, 2022); Peter Sterling and Simon Laughlin, *Principles of Neural Design* (MIT Press, 2015).

23. Ann M. Graybiel and Kyle S. Smith, "Good Habits, Bad Habits," *Scientific American* 310, no. 6 (2014): 38–43.

24. Dill and Holton, "The Addict in Us All."

25. Mitchell F. Roitman et al., "Induction of a Salt Appetite Alters Dendritic Morphology in Nucleus Accumbens and Sensitizes Rats to Amphetamine," *Journal of Neuroscience* 22, no. 11 (2002): RC225.

26. Berridge, "Affective Valence in the Brain"; Shelly B. Flagel et al., "Genetic Background and Epigenetic Modifications in the Core of the Nucleus Accumbens Predict Addiction-Like Behavior in a Rat Model," *Proceedings of the National Academy of Sciences* 113, no. 20 (2016): E2861–E2870.

27. James, *The Principles of Psychology*.

28. Engel, Friston, and Kragic, *The Pragmatic Turn*; Huebner and Schulkin, *Biological Cognition*.

29. George Koob and Mary Jeanne Kreek, "Stress, Dysregulation of Drug Reward Pathways, and the Transition to Drug Dependence," *American Journal of Psychiatry* 164, no. 8 (2007): 1149–59.

30. Ann M. Graybiel, "The Basal Ganglia and Chunking of Action Repertoires," *Neurobiology of Learning and Memory* 70, nos. 1–2 (1998): 119–36; Smith and Graybiel, "Habit Formation"; Kent C. Berridge, "Is Addiction a Brain Disease?," *Neuroethics* 10 (2017): 29–33.

31. See Huebner and Schulkin, *Biological Cognition*, for further discussion.

32. Kent C. Berridge, "From Prediction Error to Incentive Salience: Mesolimbic Computation of Reward Motivation," *European Journal of Neuroscience* 35, no. 7 (2012): 1124–43; Berridge, "Affective Valence in the Brain."

33. Huijeong Jeong et al., "Mesolimbic Dopamine Release Conveys Causal Associations," *Science* 378, no. 6626 (2022): eabq6740; Melissa J. Sharpe et al., "Dopamine Transients Are Sufficient and Necessary for Acquisition of Model-Based Associations," *Nature Neuroscience* 20, no. 5 (2017): 735–42.

34. Nora D. Volkow, Michael Michaelides, and Ruben Baler, "The Neuroscience of Drug Reward and Addiction," *Physiological Reviews* 99, no. 4 (2019): 2115–40; Berridge, "Affective Valence in the Brain."

35. George F. Koob and Michel Le Moal, *Neurobiology of Addiction* (Elsevier, 2006).

36. Kent C. Berridge, "Is Addiction a Brain Disease? The Incentive-Sensitization View," in *Evaluating the Brain Disease Model of Addiction* (Routledge, 2022), 74–86.

37. Koob and Le Moal, *Neurobiology of Addiction*.

38. Adele Diamond, "Consequences of Variations in Genes That Affect Dopamine in Prefrontal Cortex," *Cerebral Cortex* 17, no. suppl_1 (2007): i161–i170.

39. Jerome Kagan, *Surprise, Uncertainty, and Mental Structures* (Harvard University Press, 2002).

40. Everitt and Robbins, "Drug Addiction."

41. George Ainslie and John Monterosso, "A Marketplace in the Brain?," *Science* 306, no. 5695 (2004): 421–23.

42. George Ainslie, *Breakdown of Will* (Cambridge University Press, 2001); Gregory S. Berns, David Laibson, and George Loewenstein, "Intertemporal Choice—Toward an Integrative Framework," *Trends in Cognitive Sciences* 11, no. 11 (2007): 482–88.

43. Daniel Kahneman, Paul Slovic, and Amos Tversky, *Judgment Under Uncertainty: Heuristics and Biases* (Cambridge University Press, 1982).

44. Daniel Kahneman, *Thinking, Fast and Slow* (Farrar, Straus and Giroux, 2011); Gerd Gigerenzer, *Adaptive Thinking* (Oxford University Press, 2000).

45. George Loewenstein, "A Visceral Account of Addiction," in *Getting Hooked: Rationality and Addiction*, ed. Jon Elster and Ole-Jørgen Skog (Cambridge University Press, 1999), 237–45.

46. Elster, *Ulysses Unbound.*

47. Elster, *Ulysses Unbound.*

48. Graybiel and Smith, "Good Habits, Bad Habits."

49. Nayla N. Chaijale et al., "Social Stress Engages Opioid Regulation of Locus Coeruleus Norepinephrine Neurons and Induces a State of Cellular and Physical Opiate Dependence," *Neuropsychopharmacology* 38, no. 10 (2013): 1833–43; Joseph E. LeDoux, *Anxious: Using the Brain to Understand and Treat Fear and Anxiety* (Viking, 2015).

50. Ainslie and Monterosso, "A Marketplace in the Brain?"

51. Giodorno de Guglielmo et al., "Dopamine D3 Receptor Antagonism Reverses the Escalation of Oxycodone Self-Administration and

Decreases Withdrawal-Induced Hyperalgesia and Irritability-Like Behavior in Oxycodone-Dependent Heterogeneous Stock Rats," *Frontiers in Behavioral Neuroscience* 13 (2020): 292.

52. Linda Farris Kurtz and Michael Fisher, "Participation in Community Life by AA and NA Members," *Contemporary Drug Problems* 30, no. 4 (2003): 875–904.

53. Elinor Ostrom, *Governing the Commons: The Evolution of Institutions for Collective Action* (Cambridge University Press, 1990).

54. See Bryce Huebner, "Planning and Prefigurative Practice," in *The Philosophy of Daniel Dennett* (Oxford University Press, 2017), 295–327.

55. Winfried Sweet Black, *Dope: The Story of the Living Dead* (Star Company, 1928).

56. Eduard Levinstein, *Morbid Craving for Morphia: Die Morphiumsucht* (Smith, Elder, 1878).

57. Rebecca M. Todd and Maria G Manaligod, "Implicit Guidance of Attention: The Priority State Space Framework," *Cortex* 102 (2018): 121–38.

58. Elster, *Ulysses Unbound*.

59. Robert C. Neville, *The Cosmology of Freedom* (Yale University Press, 1974); Jay Schulkin, "A Pragmatist Perspective on Brains, Trust, and Choice," *Journal of Speculative Philosophy* 37, no. 1 (2023): 61–80.

60. Johann Hari, *Lost Connections: Why You're Depressed and How to Find Hope* (Bloomsbury, 2019).

61. Śāntideva, *Bodhicaryāvatāra*, trans. Kate Crosby and Andrew Skilton (Oxford University Press, 2008); Benedict de Spinoza, "Ethics," in *Spinoza: Complete Works*, trans. Samuel Shirley, ed. Michael Morgan (Hackett, 1677/2002), 213–382.

2. CRAVINGS AND OTHER MOTIVATIONS

1. Fleur L. Strand, *Neuropeptides: Regulators of Physiological Processes* (MIT Press, 1999).

2. Joe Herbert and Jay Schulkin, "Neurochemical Coding of Adaptive Responses in the Limbic System," in *Hormones, Brain and Behavior* (Academic Press, 2002), 659–89.

3. Avram Goldstein, "Opioid Peptides Endorphins in Pituitary and Brain: Studies on Opiate Receptors Have Led to Identification of

Endogenous Peptides with Morphine-Like Actions," *Science* 193, no. 4258 (1976): 1081–86; J. Hughes et al., "Identification of Two Related Pentapeptides from the Brain with Potent Opiate Agonist Activity," *Nature* 258, no. 5536 (1975): 577–79; Eric J. Simon, Jacob M. Hiller, and Irit Edelman, "Stereospecific Binding of the Potent Narcotic Analgesic [3H] Etorphine to Rat-Brain Homogenate," *Proceedings of the National Academy of Sciences* 70, no. 7 (1973): 1947–49.

4. Candace B. Pert and Solomon H. Snyder, "Opiate Receptor: Demonstration in Nervous Tissue," *Science* 179, no. 4077 (1973): 1011–14.

5. Candace B. Pert, *Molecules of Emotion: The Science Behind Mind-Body Medicine* (Scribner, 1997); Jeff Goldberg, *Anatomy of a Scientific Discovery* (Bantam, 1988).

6. Goldberg, *Anatomy of a Scientific Discovery.*

7. Christoph Stein, "Opioid Receptors," *Annual Review of Medicine* 67, no. 1 (2016): 433–51; Rita J. Valentino and Nora D. Volkow, "Untangling the Complexity of Opioid Receptor Function," *Neuropsychopharmacology* 43, no. 13 (2018): 2514–20.

8. Charles Chavkin, "Dynorphin–Still an Extraordinarily Potent Opioid Peptide," *Molecular Pharmacology* 83, no. 4 (2013): 729–36.

9. Feng Zhang et al., "Multimodal Fast Optical Interrogation of Neural Circuitry," *Nature* 446, no. 7136 (2007): 633–39.

10. Jacqueline F. McGinty, Derek van der Kooy, and Floyd E. Bloom, "The Distribution and Morphology of Opioid Peptide Immunoreactive Neurons in the Cerebral Cortex of Rats," *Journal of Neuroscience* 4, no. 4 (1984): 1104–17; Valentino and Volkow, "Untangling the Complexity of Opioid Receptor Function."

11. Susanne Dreborg et al., "Evolution of Vertebrate Opioid Receptors," *Proceedings of the National Academy of Sciences* 105, no. 40 (2008): 15487–92; Dan Larhammar, Christina Bergqvist, and Görel Sundström, "Ancestral Vertebrate Complexity of the Opioid System," *Vitamins & Hormones* 97 (2015): 95–122.

12. Taylor E. Hinchliffe and Ying Xia, "Evolutionary Distribution of the δ-Opioid Receptor: From Invertebrates to Humans," *Neural Functions of the Delta-Opioid Receptor* (2015): 67–87.

13. George F. Koob and Floyd E. Bloom, "Cellular and Molecular Mechanisms of Drug Dependence," *Science* 242, no. 4879 (1988): 715–23.

14. Alan R. Gintzler, Michael D. Gershon, and Sydney Spector, "A Non-peptide Morphine-Like Compound: Immunocytochemical Localization in the Mouse Brain," *Science* 199, no. 4327 (1978): 447–48; George Stefano et al., "Endogenous Morphine: Up-to-Date Review 2011," *Folia biologica* 58, no. 2 (2012): 49.

15. Kent C. Berridge, "Is Addiction a Brain Disease? The Incentive-Sensitization View," in *Evaluating the Brain Disease Model of Addiction* (Routledge, 2022), 74–86; Kent C. Berridge and Terry E. Robinson, "What Is the Role of Dopamine in Reward: Hedonic Impact, Reward Learning, or Incentive Salience?," *Brain Research Reviews* 28, no. 3 (1998): 309–69; Jay Schulkin, *Effort: A Behavioral Neuroscience Perspective on the Will* (Erlbaum, 2007).

16. Bruce S. McEwen, "Allostasis and the Epigenetics of Brain and Body Health over the Life Course: The Brain on Stress," *JAMA Psychiatry* 74, no. 6 (2017): 551–52.

17. John Hughes, "Isolation of an Endogenous Compound from the Brain with Pharmacological Properties Similar to Morphine," *Brain Research* 88, no. 2 (1975): 295–308; Gavril W. Pasternak and Solomon H. Snyder, "Opiate Receptor Binding: Enzymatic Treatments That Discriminate Between Agonist and Antagonist Interactions," *Molecular Pharmacology* 11, no. 4 (1975): 478–84; Lars Terenius, "Stereospecific Interaction Between Narcotic Analgesics and a Synaptic Plasma Membrane Fraction of Rat Cerebral Cortex," *Acta pharmacologica et toxicologica* 32, nos. 3–4 (1973): 317–20; Lars Terenius and Annika Wahlström, "Search for an Endogenous Ligand for the Opiate Receptor," *Acta Physiologica Scandinavica* 94, no. 1 (1975): 74–81.

18. Charles Chavkin and George F. Koob, "Dynorphin, Dysphoria, and Dependence: The Stress of Addiction," *Neuropsychopharmacology* 41, no. 1 (2016): 373.

19. Mary Jeanne Kreek, Brian Reed, and Eduardo Butelman, "Current Status of Opioid Addiction Treatment and Related Preclinical Research," *Science Advances* 5, no. 10 (2019): eaax9140.

20. For example, Sophia C. Levis et al., "On the Early Life Origins of Vulnerability to Opioid Addiction," *Molecular Psychiatry* 26, no. 8 (2021): 4409–16.

21. See George F. Koob, "The Dark Side of Emotion: The Addiction Perspective," *European Journal of Pharmacology* 753 (2015): 73–87; Michael R.

Bruchas et al., "CRF1-R Activation of the Dynorphin/Kappa Opioid System in the Mouse Basolateral Amygdala Mediates Anxiety-Like Behavior," *PloS One* 4, no. 12 (2009): e8528; Michael R. Bruchas, Benjamin B. Land, and Charles Chavkin, "The Dynorphin/Kappa Opioid System as a Modulator of Stress-Induced and Pro-Addictive Behaviors," *Brain Research* 1314 (2010): 44–55; Chavkin and Koob, "Dynorphin, Dysphoria, and Dependence."

22. Mary Jeanne Kreek et al., "Opiate and Cocaine Addiction: From Bench to Clinic and Back to the Bench," *Current Opinion in Pharmacology* 9, no. 1 (2009): 74–80; Bruchas, Land, and Chavkin, "The Dynorphin/Kappa Opioid System."

23. Solomon N. Snyder, *Brainstorming: The Science and Politics of Opiate Research* (Harvard University Press, 1989); Gavril W. Pasternak and Ying-Xian Pan, "Mu Opioids and Their Receptors: Evolution of a Concept," *Pharmacological Reviews* 65, no. 4 (2013): 1257–1317.

24. Nancy D. Campbell, *OD: Naloxone and the Politics of Overdose* (MIT Press, 2020).

25. George F. Koob and Michel Le Moal, *Neurobiology of Addiction* (Elsevier, 2006).

26. Kreek, Reed, and Butelman, "Current Status of Opioid Addiction Treatment."

27. Pasternak and Pan, "Mu Opioids and Their Receptors."

28. Batoul Sadat Haerian and Monir Sadat Haerian, "OPRM1 rs1799971 Polymorphism and Opioid Dependence: Evidence from a Meta-Analysis," *Pharmacogenomics* 14, no. 7 (2013): 813–24.

29. Pasternak and Pan, "Mu Opioids and Their Receptors."

30. Matthew T. Birnie et al., "Plasticity of the Reward Circuitry After Early-Life Adversity: Mechanisms and Significance," *Biological Psychiatry* 87, no. 10 (2020): 875–84.

31. Michael Davis, "Are Different Parts of the Extended Amygdala Involved in Fear Versus Anxiety?," *Biological Psychiatry* 44, no. 12 (1998): 1239–47; Philip W. Gold, "The Organization of the Stress System and Its Dysregulation in Depressive Illness," *Molecular Psychiatry* 20, no. 1 (2015): 32–47; Benedict J. Kolber et al., "Transient Early-Life Forebrain Corticotropin-Releasing Hormone Elevation Causes Long-Lasting Anxiogenic and Despair-Like Changes in Mice," *Journal of Neuroscience* 30, no. 7 (2010): 2571–81; Charles B. Nemeroff et al., "Elevated

Concentrations of CSF Corticotropin-Releasing Factor-Like Immu-noreactivity in Depressed Patients," *Science* 226, no. 4680 (1984): 1342–44; Jeffrey B. Rosen and Jay Schulkin, "From Normal Fear to Pathological Anxiety," *Psychological Review* 105, no. 2 (1998): 325; R. Parrish Waters et al., "Evidence for the Role of Corticotropin-Releasing Factor in Major Depressive Disorder," *Neuroscience & Biobehavioral Reviews* 58 (2015): 63–78.

32. Matthew B. Pomrenze et al., "Dissecting the Roles of GABA and Neu-ropeptides from Rat Central Amygdala CRF Neurons in Anxiety and Fear Learning," *Cell Reports* 29, no. 1 (2019): 13–21; Arun Asok et al., "Optogenetic Silencing of a Corticotropin-Releasing Factor Pathway from the Central Amygdala to the Bed Nucleus of the Stria Terminalis Disrupts Sustained Fear," *Molecular Psychiatry* 23, no. 4 (2018): 914–22.

33. Kolber et al., "Transient Early-Life Forebrain"; Mate Toth et al., "Over-expression of Forebrain CRH During Early Life Increases Trauma Susceptibility in Adulthood," *Neuropsychopharmacology* 41, no. 6 (2016): 1681–90.

34. Wallace Craig, "Appetites and Aversions as Constituents of Instincts," *Biological Bulletin* 34, no. 2 (1918): 91–107; John Dewey, *Experience and Nature* (Dover, 1925/1989).

35. Paul Rozin and Jay Schulkin, "Food Selection," in *Neurobiology of Food and Fluid Intake*, ed. Eduard M. Stricker (Plenum, 1990), 297–328.

36. Vincent Gaston Dethier, *The Hungry Fly* (Harvard University Press, 1976); Lars Chittka, *The Mind of a Bee* (Princeton University Press, 2022).

37. John Garcia and Robert A. Koelling, "Relation of Cue to Conse-quence in Avoidance Learning," *Psychonomic Science* 4 (1966): 123–24; Paul Rozin, "Specific Aversions as a Component of Specific Hungers," *Journal of Comparative and Physiological Psychology* 64, no. 2 (1967): 237; Paul Rozin, "The Evolution of Intelligence and Access to the Cognitive Unconscious," in *Progress in Psychobiology and Physiological Psychology*, ed. James M. Sprague and Alan N. Epstein (Academic Press, 1976), 245–80.

38. Daniel H. Janzen, "Why Fruits Rot, Seeds Mold, and Meat Spoils," *American Naturalist* 111, no. 980 (1977): 691–713.

39. Michael L. Power and Jay Schulkin, *The Evolution of Obesity* (Johns Hopkins University Press, 2009).

40. Curt P. Richter, "Total Self-Regulatory Functions in Animals and Human Beings," *Harvey Lecture Series*, 38, no. 63 (1943); Rozin, "The Evolution of Intelligence and Access to the Cognitive Unconscious."

41. James T. Fitzsimons, *The Physiology of Thirst and Sodium Appetite*, monographs of the Physiological Society, no. 35 (Oxford University Press, 1979); James T. Fitzsimons, "Angiotensin, Thirst, and Sodium Appetite," *Physiological Reviews* 78, no. 3 (1998): 583–686.

42. Derek Denton, *The Hunger for Salt* (Springer Verlag, 1983).

43. Steven J. Fluharty and Alan N. Epstein, "Sodium Appetite Elicited by Intracerebroventricular Infusion of Angiotensin II in the Rat: II. Synergistic Interaction with Systemic Mineralocorticoids," *Behavioral Neuroscience* 97, no. 5 (1983): 746.

44. Alan N. Epstein, "Mineralocorticoids and Cerebral Angiotensin May Act Together to Produce Sodium Appetite," *Peptides* 3, no. 3 (1982): 493–94.

45. Alan C. Spector, "Linking Gustatory Neurobiology to Behavior in Vertebrates," *Neuroscience & Biobehavioral Reviews* 24, no. 4 (2000): 391–416.

46. Derek Daniels and Jay Schulkin, "Water and Salt Intake in Vertebrates: Endocrine and Behavioral Regulation," in *Encyclopaedia of Animal Behavior* (Academic Press, 2010), 569–79; E. E. Krieckhaus, "'Innate Recognition' Aids Rats in Sodium Regulation," *Journal of Comparative and Physiological Psychology* 73, no. 1 (1970): 117; Curt P. Richter, "Salt Appetite of Mammals: Its Dependence on Instinct and Metabolism," *L'instinct dans le comportement des animaux et de l'homme. Paris* (1956): 577–629; Alan M. Rosenwasser, Jay Schulkin, and Norman T. Adler, "Anticipatory Appetitive Behavior of Adrenalectomized Rats Under Circadian Salt-Access Schedules," *Animal Learning & Behavior* 16, no. 3 (1988): 324–29; George Wolf, "Innate Mechanisms for Regulation of Sodium Intake," *Olfaction and Taste*, no. 3 (1969): 548–53.

47. Denton, *The Hunger for Salt*; Jay Schulkin, *Sodium Hunger: The Search for a Salty Taste* (Cambridge University Press, 1991); Fitzsimons, *The Physiology of Thirst and Sodium Appetite*; Fitzsimons, "Angiotensin, Thirst, and Sodium Appetite."

48. Kent C. Berridge et al., "Sodium Depletion Enhances Salt Palatability in Rats," *Behavioral Neuroscience* 98, no. 4 (1984): 652.

49. Rozin, "The Evolution of Intelligence and Access to the Cognitive Unconscious"; Rozin and Schulkin, "Food Selection."

50. See Richter, "Total Self-Regulatory Functions in Animals and Human Beings"; Andrew Scull and Jay Schulkin, "Psychobiology, Psychiatry, and Psychoanalysis: The Intersecting Careers of Adolf Meyer, Phyllis Greenacre, and Curt Richter," *Medical History* 53, no. 1 (2009): 5–36; Jay Schulkin, *Curt Richter: A Life in the Laboratory* (Johns Hopkins University Press, 2005).

51. Michael Bliss, "Resurrections in Toronto: Fact and Myth in the Discovery of Insulin," *Bulletin of the American Academy of Arts and Sciences* 38, no. 3 (1984): 15–36; David P. Todes, *Pavlov's Physiology Factory: Experiment, Interpretation, Laboratory Enterprise* (Johns Hopkins University Press, 2002).

52. Ivan P. Pavlov, *The Work of the Digestive Glands* (Charles Griffin, 1897/1902); Ivan P. Pavlov, *Lectures on Conditioned Reflexes* (International Publishing, 1927/1960).

53. Terry L. Powley, "The Ventromedial Hypothalamic Syndrome, Satiety, and a Cephalic Phase Hypothesis," *Psychological Review* 84, no. 1 (1977): 89.

54. Power and Schulkin, *The Evolution of Obesity*.

55. Power and Schulkin, *The Evolution of Obesity*.

56. Randolph M. Nesse and Kent C. Berridge, "Psychoactive Drug Use in Evolutionary Perspective," *Science* 278, no. 5335 (1997): 63–66; Euclid O. Smith, "Evolution, Substance Abuse and Addiction," in *Evolutionary Medicine*, ed. Wenda R. Trevathan (Oxford University Press, 1999), 375–406.

57. Sharif A. Taha et al., "Endogenous Opioids Encode Relative Taste Preference," *European Journal of Neuroscience* 24, no. 4 (2006): 1220–26.

58. Joel Fort, *The Pleasure Seekers: The Drug Crisis, Youth, and Society* (Bobbs-Merrill, 1969).

59. Jerome Kagan, *Surprise, Uncertainty, and Mental Structures* (Harvard University Press, 2002).

60. Charles P. O'Brien et al., "Penn/VA Center for Studies of Addiction," *Neuropharmacology* 56 (2009): 44–47.

61. See Daniel Kahneman, *Thinking, Fast and Slow* (Farrar, Straus and Giroux, 2011).

62. James R. Stellar and Eliot Stellar, "Physiological Aspects of Motivation and Reward," *Neurobiology of Motivation and Reward* (1985): 51–82.

63. Rozin, "The Evolution of Intelligence and Access to the Cognitive Unconscious"

64. Debra A. Zellner et al., "Rats Learn to Like the Taste of Morphine," *Behavioral Neuroscience* 99, no. 2 (1985): 290.

65. Berridge et al., "Sodium Depletion Enhances Salt Palatability in Rats."

66. Michael N. Nitabach, Jay Schulkin, and Alan N. Epstein, "The Medial Amygdala Is Part of a Mineralocorticoid-Sensitive Circuit Controlling NaCl Intake in the Rat," *Behavioural Brain Research* 35, no. 2 (1989): 127–34; Randy J. Seeley et al., "Lesions of the Central Nucleus of the Amygdala II: Effects on Intraoral NaCl Intake," *Behavioural Brain Research* 59, nos. 1–2 (1993): 19–25.

67. Craig M. Smith et al., "Endogenous Central Amygdala Mu-Opioid Receptor Signaling Promotes Sodium Appetite in Mice," *Proceedings of the National Academy of Sciences* 113, no. 48 (2016): 13893–98.

68. Francis W. Flynn et al., "Central Gustatory Lesions: II. Effects on Sodium Appetite, Taste Aversion Learning, and Feeding Behaviors," *Behavioral Neuroscience* 105, no. 6 (1991): 944; Spector, "Linking Gustatory Neurobiology to Behavior in Vertebrates."

69. Lisandra Brandino de Oliveira, Laurival A. De Luca Jr., and José Vanderlei Menani, "Opioid Activation in the Lateral Parabrachial Nucleus Induces Hypertonic Sodium Intake," *Neuroscience* 155, no. 2 (2008): 350–58; Junbao Yan et al., "Activation of μ-Opioid Receptors in the Central Nucleus of the Amygdala Induces Hypertonic Sodium Intake," *Neuroscience* 233 (2013): 28–43.

70. Robert F. Lundy Jr. and Ralph Norgren, "Gustatory System," in *The Rat Nervous System* (Academic Press, 2004), 890–920; Elisa S. Na, Michael J. Morris, and Alan Kim Johnson, "Opioid Mechanisms That Mediate the Palatability of and Appetite for Salt in Sodium Replete and Deficient States," *Physiology & Behavior* 106, no. 2 (2012): 164–70.

71. Mary Bertino et al., "A Small Dose of Morphine Increases Intake of and Preference for Isotonic Saline Among Rats," *Pharmacology Biochemistry and Behavior* 29, no. 3 (1988): 617–23; Sharon Clarke and Linda A. Parker, "Morphine-Induced Modification of Quinine Palatability: Effects of Multiple Morphine-Quinine Trials," *Pharmacology Biochemistry and Behavior* 51, nos. 2–3 (1995): 505–8.

72. Richard J. Bodnar, "Endogenous Opiates and Behavior," *Peptides* 75 (2016): 18–70; Richard J. Bodnar, Michael J. Glass, and James E. Koch, "Analysis of Central Opioid Receptor Subtype Antagonism of Hypotonic and Hypertonic Saline Intake in Water-Deprived Rats," *Brain Research Bulletin* 36, no. 3 (1995): 293–300.

73. Predrag Petrovic et al., "Blocking Central Opiate Function Modulates Hedonic Impact and Anterior Cingulate Response to Rewards and Losses," *Journal of Neuroscience* 28, no. 42 (2008): 10509–16.

74. Schulkin, *Sodium Hunger*; Jay Schulkin and Peter Sterling, "Allostasis: A Brain-Centered, Predictive Mode of Physiological Regulation," *Trends in Neurosciences* 42, no. 10 (2019): 740–52; Spector, "Linking Gustatory Neurobiology to Behavior in Vertebrates."

75. Anthony Dickinson and Bernard Balleine, "Motivational Control of Goal-Directed Action," *Animal Learning & Behavior* 22, no. 1 (1994): 1–18.; E. E. Krieckhaus and George Wolf, "Acquisition of Sodium by Rats: Interaction of Innate Mechanisms and Latent Learning," *Journal of Comparative and Physiological Psychology* 65, no. 2 (1968): 197; Kent C. Berridge and Jay Schulkin, "Palatability Shift of a Salt-Associated Incentive During Sodium Depletion," *Quarterly Journal of Experimental Psychology* 41, no. 2 (1989): 121–38.

76. Henning Boecker et al., "The Runner's High: Opioidergic Mechanisms in the Human Brain," *Cerebral Cortex* 18, no. 11 (2008): 2523–31.

77. Susana Peciña and Kent C. Berridge, "Dopamine or Opioid Stimulation of Nucleus Accumbens Similarly Amplify Cue-Triggered 'Wanting' for Reward: Entire Core and Medial Shell Mapped as Substrates for PIT Enhancement," *European Journal of Neuroscience* 37, no. 9 (2013): 1529–40; Susana Peciña, Kyle S. Smith, and Kent C. Berridge, "Hedonic Hot Spots in the Brain," *Neuroscientist* 12, no. 6 (2006): 500–511; Daniel C. Castro and Kent C. Berridge, "Opioid Hedonic Hotspot in Nucleus Accumbens Shell: Mu, Delta, and Kappa Maps for Enhancement of Sweetness 'Liking' and 'Wanting,'" *Journal of Neuroscience* 34, no. 12 (2014): 4239–50; Craig M. Smith and Andrew J. Lawrence, "Salt Appetite, and the Influence of Opioids," *Neurochemical Research* 43, no. 1 (2018): 12–18.

78. Koob and Le Moal, *Neurobiology of Addiction*.

79. George F. Koob, "Drug Addiction: Hyperkatifeia/Negative Reinforcement as a Framework for Medications Development," *Pharmacological Reviews* 73, no. 1 (2021): 163–201.

80. Peter Sterling, *What Is Health? Allostasis and the Evolution of Human Design* (MIT Press, 2020).

81. Randolph M. Nesse and Jay Schulkin, "An Evolutionary Medicine Perspective on Pain and Its Disorders," *Philosophical Transactions of the Royal Society B* 374, no. 1785 (2019): 20190288; Edgar T. Walters and Amanda C. de Williams, "Evolution of Mechanisms and Behaviour Important for Pain," *Philosophical Transactions of the Royal Society B* 374, no. 1785 (2019): 20190275.

82. Nesse and Schulkin, "An Evolutionary Medicine Perspective on Pain and Its Disorders."

83. Richard Ambron, *The Brain and Pain: Breakthroughs in Neuroscience* (Columbia University Press, 2022); Gregory Corder et al., "Endogenous and Exogenous Opioids in Pain," *Annual Review of Neuroscience* 41, no. 1 (2018): 453–73.

84. Ronald Melzack and Patrick D. Wall, "Pain Mechanisms: A New Theory," *Science* 150, no. 3699 (1965): 971–79.

85. Colin Klein, *What the Body Commands: The Imperative Theory of Pain* (MIT Press, 2015).

86. Nitabach unpublished observations.

87. Linda R. Watkins et al., "Neurocircuitry of Illness-Induced Hyperalgesia," *Brain Research* 639, no. 2 (1994): 283–99; Steven F. Maier, Eric P. Wiertelak, and Linda R. Watkins, "Endogenous Pain Facilitory Systems Antianalgesia and Hyperalgesia," *APS Journal* 1, no. 3 (1992): 191–98.

88. Nesse and Schulkin, "An Evolutionary Medicine Perspective on Pain and Its Disorders."

89. Charles Darwin, *The Life and Letters of Charles Darwin: Including an Autobiographical Chapter*, vol. 1 (D. Appleton, 1887), 51–52.

90. Amaury François et al., "A Brainstem-Spinal Cord Inhibitory Circuit for Mechanical Pain Modulation by GABA and Enkephalins," *Neuron* 93, no. 4 (2017): 822–39.

91. Klein, *What the Body Commands*; Nesse and Schulkin, "An Evolutionary Medicine Perspective on Pain and Its Disorders."

92. Koob, "Drug Addiction."

93. Roy Porter, *The Greatest Benefit to Mankind: A Medical History of Humanity*, Norton History of Science (Norton, 1997).

94. Koob, "Drug Addiction."

95. Campbell, *OD*.
96. Klein, *What the Body Commands*.
97. Campbell, *OD*.
98. Nora D. Volkow, George F. Koob, and A. Thomas McLellan, "Neurobiologic Advances from the Brain Disease Model of Addiction," *New England Journal of Medicine* 374, no. 4 (2016): 363–71.
99. Campbell, *OD*.
100. Carl L. Hart, *High Price: A Neuroscientist's Journey of Self-Discovery That Challenges Everything You Know About Drugs and Society* (Harper, 2013); Carl L. Hart, *Drug Use for Grown-Ups: Chasing Liberty in the Land of Fear* (Penguin, 2021).
101. David Herzberg, *White Market Drugs: Big Pharma and the Hidden History of Addiction in America* (University of Chicago Press, 2020).
102. Claire D. Clark, *The Recovery Revolution: The Battle over Addiction Treatment in the United States* (Columbia University Press, 2017), 210.
103. Schulkin, *Sodium Hunger*.
104. Thomas Davison Crothers, *Morphinism and Narcomanias from Other Drugs: Their Etiology, Treatment, and Medicolegal Relations* (W. B. Saunders, 1902), 33.
105. See Markus Heilig, *The Thirteenth Step: Addiction in the Age of Brain Science* (Columbia University Press, 2015).
106. Crothers, *Morphinism and Narcomanias from Other Drugs*, 98.
107. Bryce Huebner and Jay Schulkin, *Biological Cognition* (Cambridge University Press, 2022); Peter Sterling and Simon Laughlin, *Principles of Neural Design* (MIT Press, 2015).
108. Koob and Moal, *Neurobiology of Addiction*.
109. Jeffery S. Mogil, "Sex Differences in Pain and Pain Inhibition: Multiple Explanations of a Controversial Phenomenon," *Nature Reviews Neuroscience* 13, no. 12 (2012): 859–66; Jeffrey S. Mogil and Benjamin Kest, "Sex Differences in Opioid Analgesia: Of Mice and Women," *Pain Forum* 8, no. 1 (1999): 48–50; Jeffrey S. Mogil et al., "The Genetics of Pain and Pain Inhibition," *Proceedings of the National Academy of Sciences* 93, no. 7 (1996): 3048–55.
110. Elaine Scarry, *The Body in Pain: The Making and Unmaking of the World* (Oxford University Press, 1985).

3. REGULATION: EMOTION AND ANGST

1. Jules Hirsch, "One Thing Leads to Another," *Journal of Clinical Investigation* 114, no. 8 (2004): 1040–43.

2. Vincent P. Dole, "Biochemistry of Addiction," *Annual Review of Chemistry* 39 (1970): 821–40; Vincent P. Dole, "Implications of Methadone Maintenance for Theories of Narcotic Addiction," *JAMA* 260, no. 20 (1988): 3025–29.

3. Vincent P. Dole and Marie E. Nyswander, "Heroin Addiction—A Metabolic Disease," *Archives of Internal Medicine* 120, no. 1 (1967): 19–24.

4. Julie Netherland and Helena Hansen, "White Opioids: Pharmaceutical Race and the War on Drugs That Wasn't," *BioSocieties* 12, no. 2 (2017): 217–38.

5. Jay Schulkin, *Adaptation and Well-Being: Social Allostasis* (Cambridge University Press, 2011).

6. Jay Schulkin and Peter Sterling, "Allostasis: A Brain-Centered, Predictive Mode of Physiological Regulation," *Trends in Neurosciences* 42, no. 10 (2019): 740–52.

7. Peter Sterling and Joseph Eyer, "Allostasis: A New Paradigm to Explain Arousal Pathology," in *Handbook of Life Stress, Cognition and Health*, ed. Shirley Fisher and James Reason (Wiley, 1988), 629–49; Peter Sterling, *What Is Health? Allostasis and the Evolution of Human Design* (MIT Press, 2020).

8. Bruce S. McEwen and Eliot Stellar, "Stress and the Individual: Mechanisms Leading to Disease," *Archives of Internal Medicine* 153, no. 18 (1993): 2093–2101; Bruce S. McEwen, "Stress, Adaptation, and Disease: Allostasis and Allostatic Load," *Annals of the New York Academy of Sciences* 840, no. 1 (1998): 33–44.

9. Peter Sterling and Simon Laughlin, *Principles of Neural Design* (MIT Press, 2015); Jay Schulkin, Bruce S. McEwen, and Philip W. Gold, "Allostasis, Amygdala, and Anticipatory Angst," *Neuroscience & Biobehavioral Reviews* 18, no. 3 (1994): 385–96.

10. Bryce Huebner and Jay Schulkin, *Biological Cognition* (Cambridge University Press, 2022).

11. George F. Koob and Jay Schulkin, "Addiction and Stress: An Allostatic View," *Neuroscience & Biobehavioral Reviews* 106 (2019): 245–62.

12. Sterling, *What Is Health?*

13. George F. Koob and Michel Le Moal, *Neurobiology of Addiction* (Elsevier, 2006); McEwen and Stellar, "Stress and the Individual."

14. Koob and Le Moal, *Neurobiology of Addiction*.

15. George F. Koob and Floyd E. Bloom, "Cellular and Molecular Mechanisms of Drug Dependence," *Science* 242, no. 4879 (1988): 715–23; Wolfgang B. Liedtke et al., "Relation of Addiction Genes to Hypothalamic Gene Changes Subserving Genesis and Gratification of a Classic Instinct, Sodium Appetite," *Proceedings of the National Academy of Sciences* 108, no. 30 (2011): 12509–14.

16. Koob and Le Moal, *Neurobiology of Addiction*; George Koob and Mary Jeanne Kreek, "Stress, Dysregulation of Drug Reward Pathways, and the Transition to Drug Dependence," *American Journal of Psychiatry* 164, no. 8 (2007): 1149–59; Koob and Schulkin, "Addiction and Stress."

17. Judith Grisel, *Never Enough: The Neuroscience and Experience of Addiction* (Anchor, 2019).

18. Koob and Le Moal, *Neurobiology of Addiction*; Koob and Schulkin, "Addiction and Stress."

19. Stephanie A. Carmack et al., "Heroin Addiction Engages Negative Emotional Learning Brain Circuits in Rats," *Journal of Clinical Investigation* 129, no. 6 (2019): 2480–84.

20. George F. Koob, "Drug Addiction: Hyperkatifeia/Negative Reinforcement as a Framework for Medications Development," *Pharmacological Reviews* 73, no. 1 (2021): 163–201.

21. Andreas Lüthi and Christian Lüscher, "Pathological Circuit Function Underlying Addiction and Anxiety Disorders," *Nature Neuroscience* 17, no. 12 (2014): 1635–43.

22. Charles Darwin, *The Expression of the Emotions in Man and Animals* (Oxford University Press, 1872/1998); John Dewey, "The Reflex Arc Concept in Psychology," *Psychological Review* 3, no. 4 (1896): 357.

23. Antoine Bechara, "The Role of Emotion in Decision-Making: Evidence from Neurological Patients with Orbitofrontal Damage," *Brain and Cognition* 55, no. 1 (2004): 30–40.

24. Darwin, *The Expression of the Emotions in Man and Animals*.

25. Kent C. Berridge, "Is Addiction a Brain Disease?," *Neuroethics* 10 (2017): 29–33; Huebner and Schulkin, *Biological Cognition*; Luiz Pessoa, "Embracing Integration and Complexity: Placing Emotion Within a

Science of Brain and Behaviour," *Cognition and Emotion* 33, no. 1 (2019): 55–60; Jay Schulkin, *Roots of Social Sensibility and Neural Function* (MIT Press, 2000); Jay Schulkin, *Rethinking Homeostasis* (MIT Press, 2003).

26. Bryce Huebner, "Picturing, Signifying, and Attending," *Belgrade Philosophical Annual*, no. 31 (2018): 7–40; Rebecca M. Todd et al., "Emotional Objectivity: Neural Representations of Emotions and their Interaction with Cognition," *Annual Review of Psychology* 71, no. 1 (2020): 25–48.

27. Jeffrey B. Rosen and Jay Schulkin, "From Normal Fear to Pathological Anxiety," *Psychological Review* 105, no. 2 (1998): 325.

28. Huebner and Schulkin, *Biological Cognition*.

29. Randolph M. Nesse, *Good Reasons for Bad Feelings: Insights from the Frontier of Evolutionary Psychiatry* (Penguin, 2019); Randolph M. Nesse and Kent C. Berridge, "Psychoactive Drug Use in Evolutionary Perspective," *Science* 278, no. 5335 (1997): 63–66; Euclid O. Smith, "Evolution, Substance Abuse and Addiction," in *Evolutionary Medicine*, ed. Wenda R. Trevathan (Oxford University Press, 1999), 375–406.

30. Elisabeth B. Binder and Charles B. Nemeroff, "The CRF System, Stress, Depression and Anxiety—Insights from Human Genetic Studies," *Molecular Psychiatry* 15, no. 6 (2010): 574–88; Philip W. Gold, "The Organization of the Stress System and Its Dysregulation in Depressive Illness," *Molecular Psychiatry* 20, no. 1 (2015): 32–47; Susan K. Wood et al., "Individual Differences in Reactivity to Social Stress Predict Susceptibility and Resilience to a Depressive Phenotype: Role of Corticotropin-Releasing Factor," *Endocrinology* 151, no. 4 (2010): 1795–1805.

31. Nicole M. Wanner, Mathia L. Colwell, and Christopher Faulk, "The Epigenetic Legacy of Illicit Drugs: Developmental Exposures and Late-Life Phenotypes," *Environmental Epigenetics* 5, no. 4 (2019): dvz022.

32. Ja Wook Koo et al., "Epigenetic Basis of Opiate Suppression of Bdnf Gene Expression in the Ventral Tegmental Area," *Nature Neuroscience* 18, no. 3 (2015): 415–22.

33. Koob and Schulkin, "Addiction and Stress"; Paula E. Park et al., "Chronic CRF 1 Receptor Blockade Reduces Heroin Intake Escalation and Dependence-Induced Hyperalgesia," *Addiction Biology* 20, no. 2 (2015): 275–84.

34. Artur H. Swiergiel et al., "Antagonism of Corticotropin-Releasing Factor Receptors in the Locus Coeruleus Attenuates Shock-Induced

Freezing in Rats," *Brain Research* 587, no. 2 (1992): 263–68; Kazuhiro Takahashi et al., "Corticotropin-Releasing Hormone in the Human Hypothalamus. Free-Floating Immunostaining Method," *Endocrinologia Japonica* 36, no. 2 (1989): 275–80.

35. Zul Merali et al., "Aversive and Appetitive Events Evoke the Release of Corticotropin-Releasing Hormone and Bombesin-Like Peptides at the Central Nucleus of the Amygdala," *Journal of Neuroscience* 18, no. 12 (1998): 4758–66.

36. Christina A. Sanford et al., "A Central Amygdala CRF Circuit Facilitates Learning About Weak Threats," *Neuron* 93, no. 1 (2017): 164–78.

37. Koob and Le Moal, *Neurobiology of Addiction*; Koob, "Drug Addiction."

38. Jay Schulkin, *The CRF Signal* (Oxford University Press, 2017).

39. Eva Tarjan, Derek A. Denton, and Richard S. Weisinger, "Corticotropin-Releasing Factor Enhances Sodium and Water Intake/Excretion in Rabbits," *Brain Research* 542, no. 2 (1991): 219–24; Eva Tarjan and Derek A. Denton, "Sodium/Water Intake of Rabbits Following Administration of Hormones of Stress," *Brain Research Bulletin* 26, no. 1 (1991): 133–36.

40. Susana Peciña, Kyle S. Smith, and Kent C. Berridge, "Hedonic Hot Spots in the Brain," *Neuroscientist* 12, no. 6 (2006): 500–511.

41. Schulkin, *The CRF Signal*.

42. Julia C. Lemos et al., "Severe Stress Switches CRF Action in the Nucleus Accumbens from Appetitive to Aversive," *Nature* 490, no. 7420 (2012): 402–6; Susana Peciña, Jay Schulkin, and Kent C. Berridge, "Nucleus Accumbens Corticotropin-Releasing Factor Increases Cue-Triggered Motivation for Sucrose Reward: Paradoxical Positive Incentive Effects in Stress?," *BMC Biology* 4 (2006): 1–16.

43. Jean Luc Cadet et al., "Enhanced Upregulation of CRH mRNA Expression in the Nucleus Accumbens of Male Rats After a Second Injection of Methamphetamine Given Thirty Days Later," *PloS One* 9, no. 1 (2014): e84665.

44. Koob and Schulkin, "Addiction and Stress"; Mary Jeanne Kreek, Brian Reed, and Eduardo Butelman, "Current Status of Opioid Addiction Treatment and Related Preclinical Research," *Science Advances* 5, no. 10 (2019): eaax9140.

45. Hannah M. Baumgartner, Jay Schulkin, and Kent C. Berridge, "Activating Corticotropin-Releasing Factor Systems in the Nucleus Accumbens,

Amygdala, and Bed Nucleus of Stria Terminalis: Incentive Motivation or Aversive Motivation?," *Biological Psychiatry* 89, no. 12 (2021): 1162–75; Hannah M. Baumgartner et al., "Corticotropin Releasing Factor (CRF) Systems: Promoting Cocaine Pursuit Without Distress via Incentive Motivation," *PLoS One* 17, no. 5 (2022): e0267345.

46. Schulkin, McEwen, and Gold, "Allostasis, Amygdala, and Anticipatory Angst"; Larry Swanson and Donna. M. Simmons, "Differential Steroid Hormone and Neural Influences on Peptide mRNA Levels in CRH Cells of the Paraventricular Nucleus: A Hybridization Histochemical Study in the Rat," *Journal of Comparative Neurology* 285, no. 4 (1989): 413–35; Shinya Makino, Phillip W. Gold, and Jay Schulkin, "Corticosterone Effects on Corticotropin-Releasing Hormone mRNA in the Central Nucleus of the Amygdala and the Parvocellular Region of the Paraventricular Nucleus of the Hypothalamus," *Brain Research* 640, nos. 1–2 (1994): 105–12; Rosen and Schulkin, "From Normal Fear to Pathological Anxiety"; Alan G. Watts and Graciela Sanchez-Watts, "Region-Specific Regulation of Neuropeptide mRNAs in Rat Limbic Forebrain Neurones by Aldosterone and Corticosterone," *Journal of Physiology* 484, no. 3 (1995): 721–36.

47. Jason J. Radley et al., "The Contingency of Cocaine Administration Accounts for Structural and Functional Medial Prefrontal Deficits and Increased Adrenocortical Activation," *Journal of Neuroscience* 35, no. 34 (2015): 11897–910.

48. Koob and Schulkin, "Addiction and Stress."

49. Koob and Le Moal, *Neurobiology of Addiction*.

50. Michael Davis et al., "Phasic vs Sustained Fear in Rats and Humans: Role of the Extended Amygdala in Fear vs Anxiety," *Neuropsychopharmacology* 35, no. 1 (2010): 105–35.

51. Suzanne Erb et al., "A Role for the CRF-Containing Pathway from Central Nucleus of the Amygdala to Bed Nucleus of the Stria Terminalis in the Stress-Induced Reinstatement of Cocaine Seeking in Rats," *Psychopharmacology* 158 (2001): 360–65.

52. Schulkin, McEwen, and Gold, "Allostasis, Amygdala, and Anticipatory Angst."

53. Suzanne Erb and Jane Stewart, "A Role for the Bed Nucleus of the Stria Terminalis, but not the Amygdala, in the Effects of Corticotropin-Releasing

Factor on Stress-Induced Reinstatement of Cocaine Seeking," *Journal of Neuroscience* 19, no. 20 (1999): RC35; Yavin Shaham et al., "Corticotropin-Releasing Factor, but not Corticosterone, Is Involved in Stress-Induced Relapse to Heroin-Seeking in Rats," *Journal of Neuroscience* 17, no. 7 (1997): 2605–14; George F. Koob, "The Dark Side of Emotion: The Addiction Perspective," *European Journal of Pharmacology* 753 (2015): 73–87.

54. Makino, Gold, and Schulkin, "Corticosterone Effects on Corticotropin-Releasing Hormone mRNA"; Watts and Sanchez-Watts, "Region-Specific Regulation of Neuropeptide mRNAs."

55. Curt P. Richter, "Total Self-Regulatory Functions in Animals and Human Beings," *Harvey Lecture Series* 38, no. 63 (1943); George Wolf, "Innate Mechanisms for Regulation of Sodium Intake," *Olfaction and Taste*, no. 3 (1969): 548–53.

56. Christian J. Cook, "Glucocorticoid Feedback Increases the Sensitivity of the Limbic System to Stress," *Physiology & Behavior* 75, no. 4 (2002): 455–64.

57. Schulkin, *The CRF Signal.*

58. Makino, Gold, and Schulkin, "Corticosterone Effects on Corticotropin-Releasing Hormone mRNA"; Watts and Sanchez-Watts, "Region-Specific Regulation of Neuropeptide mRNAs."

59. Leandro F. Vendruscolo et al., "Corticosteroid-Dependent Plasticity Mediates Compulsive Alcohol Drinking in Rats," *Journal of Neuroscience* 32, no. 22 (2012): 7563–71.

60. Stephanie A. Carmack et al., "Corticosteroid Sensitization Drives Opioid Addiction," *Molecular Psychiatry* 27, no. 5 (2022): 2492–2501.

61. Cook, "Glucocorticoid Feedback Increases the Sensitivity of the Limbic System to Stress"; Zul Merali et al., "Effects of Corticosterone on Corticotrophin-Releasing Hormone and Gastrin-Releasing Peptide Release in Response to an Aversive Stimulus in Two Regions of the Forebrain (Central Nucleus of the Amygdala and Prefrontal Cortex)," *European Journal of Neuroscience* 28, no. 1 (2008): 165–72.

62. Bruce S. McEwen, "Protective and Damaging Effects of the Mediators of Stress and Adaptation: Allostasis and Allostatic Load," *Allostasis, Homeostasis, and the Costs of Physiological Adaptation* (2004): 65–98; Bruce S. McEwen, "Physiology and Neurobiology of Stress and Adaptation: Central Role of the Brain," *Physiological Reviews* 87, no. 3 (2007): 873–904.

63. Schulkin, *The CRF Signal.*

64. Compare Carmack et al., "Corticosteroid Sensitization Drives Opioid Addiction," 2500.

65. Carmack et al., "Corticosteroid Sensitization Drives Opioid Addiction"; Pietro Pablo Sanna et al., "11β-Hydroxysteroid Dehydrogenase Inhibition as a New Potential Therapeutic Target for Alcohol Abuse," *Translational Psychiatry* 6, no. 3 (2016): e760–e760.

66. Makino, Gold, and Schulkin, "Corticosterone Effects on Corticotropin-Releasing Hormone mRNA"; Watts and Sanchez-Watts, "Region-Specific Regulation of Neuropeptide mRNAs."

67. Carmack et al., "Corticosteroid Sensitization Drives Opioid Addiction"; Natasa Hlavacova et al., "Subchronic Treatment with Aldosterone Induces Depression-Like Behaviours and Gene Expression Changes Relevant to Major Depressive Disorder," *International Journal of Neuropsychopharmacology* 15, no. 2 (2012): 247–65; Brent Myers and Beverley Greenwood-Van Meerveld, "Divergent Effects of Amygdala Glucocorticoid and Mineralocorticoid Receptors in the Regulation of Visceral and Somatic Pain," *American Journal of Physiology-Gastrointestinal and Liver Physiology* 298, no. 2 (2010): G295–G303; Schulkin, *The CRF Signal.*

68. Koob and Schulkin, "Addiction and Stress"; Koob, "Drug Addiction."

69. McEwen and Stellar, "Stress and the Individual."

70. Bruce S. McEwen, "Allostasis and the Epigenetics of Brain and Body Health over the Life Course: The Brain on Stress," *JAMA Psychiatry* 74, no. 6 (2017): 551–52.

71. Mary Jeanne Kreek et al., "Opiate and Cocaine Addiction: From Bench to Clinic and Back to the Bench," *Current Opinion in Pharmacology* 9, no. 1 (2009): 74–80.

72. McEwen, "Allostasis and the Epigenetics of Brain and Body Health."

73. Jay Schulkin, Maria A. Morgan, and Jeffrey B. Rosen, "A Neuroendocrine Mechanism for Sustaining Fear," *Trends in Neurosciences* 28, no. 12 (2005): 629–35; Schulkin, *The CRF Signal.*

74. Cook, "Glucocorticoid Feedback Increases the Sensitivity of the Limbic System to Stress"; J. Megan Gray et al., "Sustained Glucocorticoid Exposure Recruits Cortico-Limbic CRH Signaling to Modulate Endocannabinoid Function," *Psychoneuroendocrinology* 66 (2016): 151–58; Makino, Gold, and Schulkin, "Corticosterone Effects on Corticotropin-

Releasing Hormone mRNA"; Merali et al., "Effects of Corticosterone on Corticotrophin-Releasing Hormone"; Watts and Sanchez-Watts, "Region-Specific Regulation of Neuropeptide mRNAs."

75. Koob and Le Moal, *Neurobiology of Addiction*; George F. Koob and Eric P. Zorrilla, "Neurobiological Mechanisms of Addiction: Focus on Corticotropin-Releasing Factor," *Current Opinion in Investigational Drugs* 11, no. 1 (2010): 63.

76. Koob and Le Moal, *Neurobiology of Addiction*.

77. Koob and Le Moal, *Neurobiology of Addiction*.

78. Adrie W. Bruijnzeel, "Kappa-Opioid Receptor Signaling and Brain Reward Function," *Brain Research Reviews* 62, no. 1 (2009): 127–46.

79. Juan Zhao et al., "Different Stress-Related Gene Expression in Depression and Suicide," *Journal of Psychiatric Research* 68 (2015): 176–85.

80. Koob, "The Dark Side of Emotion."

81. Ileana Morales and Kent C. Berridge, "'Liking' and 'Wanting' in Eating and Food Reward: Brain Mechanisms and Clinical Implications," *Physiology & Behavior* 227 (2020): 113152; Sara Karimi et al., "Role of Intra-Accumbal Cannabinoid CB1 Receptors in the Potentiation, Acquisition and Expression of Morphine-Induced Conditioned Place Preference," *Behavioural Brain Research* 247 (2013): 125–31.

82. Marci R. Mitchell, Kent C. Berridge, and Stephen V. Mahler, "Endocannabinoid-Enhanced 'Liking' in Nucleus Accumbens Shell Hedonic Hotspot Requires Endogenous Opioid Signals," *Cannabis and Cannabinoid Research* 3, no. 1 (2018): 166–70.

83. Matthew N. Hill et al., "Endogenous Cannabinoid Signaling Is Essential for Stress Adaptation," *Proceedings of the National Academy of Sciences* 107, no. 20 (2010): 9406–11.

84. Loren H. Parsons and Yasmin L. Hurd, "Endocannabinoid Signaling in Reward and Addiction," *Nature Reviews Neuroscience* 16, no. 10 (2015): 579–94.

85. Hill et al., "Endogenous Cannabinoid Signaling Is Essential for Stress Adaptation."

86. Susana Peciña and Kent C. Berridge, "Hedonic Hot Spot in Nucleus Accumbens Shell: Where do μ-Opioids Cause Increased Hedonic Impact of Sweetness?," *Journal of Neuroscience* 25, no. 50 (2005): 11777–86.

87. Susana Peciña and Kent C. Berridge, "Dopamine or Opioid Stimulation of Nucleus Accumbens Similarly Amplify Cue-Triggered 'Wanting' for

Reward: Entire Core and Medial Shell Mapped as Substrates for PIT Enhancement," *European Journal of Neuroscience* 37, no. 9 (2013): 1529–40.

88. R. Parrish Waters et al., "Evidence for the Role of Corticotropin-Releasing Factor in Major Depressive Disorder," *Neuroscience & Biobehavioral Reviews* 58 (2015): 63–78; Benedict J. Kolber et al., "Transient Early-Life Forebrain Corticotropin-Releasing Hormone Elevation Causes Long-Lasting Anxiogenic and Despair-Like Changes in Mice," *Journal of Neuroscience* 30, no. 7 (2010): 2571–81.

89. Nathan J. Marchant, Valerie S. Densmore, and Peregrine B. Osborne, "Coexpression of Prodynorphin and Corticotrophin-Releasing Hormone in the Rat Central Amygdala: Evidence of Two Distinct Endogenous Opioid Systems in the Lateral Division," *Journal of Comparative Neurology* 504, no. 6 (2007): 702–15; Marlene A. Wilson and Alexander J. McDonald, "The Amygdalar Opioid System," *Handbook of Behavioral Neuroscience* 26 (2020): 161–212.

90. Gorica D. Petrovich et al., "Associative Fear Conditioning of Enkephalin mRNA Levels in Central Amygdalar Neurons," *Behavioral Neuroscience* 114, no. 4 (2000): 681.

91. Nicole A. Crowley et al., "Dynorphin Controls the Gain of an Amygdalar Anxiety Circuit," *Cell Reports* 14, no. 12 (2016): 2774–83.

92. Koob and Le Moal, *Neurobiology of Addiction*; Koob, "Drug Addiction."

93. Michael R. Bruchas et al., "CRF1-R Activation of the Dynorphin/Kappa Opioid System in the Mouse Basolateral Amygdala Mediates Anxiety-Like Behavior," *PloS One* 4, no. 12 (2009): e8528; Park et al., "Chronic CRF 1 Receptor Blockade Reduces Heroin Intake Escalation"; Francesco Papaleo, Pierre Kitchener, and Angelo Contarino, "Disruption of the CRF/CRF1 Receptor Stress System Exacerbates the Somatic Signs of Opiate Withdrawal," *Neuron* 53, no. 4 (2007): 577–89; Kelly H. Skelton et al., "The CRF1 Receptor Antagonist, R121919, Attenuates the Severity of Precipitated Morphine Withdrawal," *European Journal of Pharmacology* 571, no. 1 (2007): 17–24; Marcus M. Weera et al., "Generation of a CRF1-Cre Transgenic Rat and the Role of Central Amygdala CRF1 Cells in Nociception and Anxiety-Like Behavior," *Elife* 11 (2022): e67822.

94. James Olds and Peter Milner, "Positive Reinforcement Produced by Electrical Stimulation of Septal Area and Other Regions of Rat Brain," *Journal of Comparative and Physiological Psychology* 47, no. 6 (1954): 419; Elliot S. Valenstein, *Brain Control* (Wiley, 1973).

95. Roy A. Wise, "Neurobiology of Addiction," *Current Opinion in Neurobiology* 6, no. 2 (1996): 243–51.

96. Cecily M. Whiteley, "Depression as a Disorder of Consciousness," *British Journal for the Philosophy of Science* (2021).

97. Aaron T. Beck, *Depression: Clinical, Experimental and Theoretical Aspects* (Harper & Row, 1967); Zachary D. Cohen and Robert J. DeRubeis, "Treatment Selection in Depression," *Annual Review of Clinical Psychology* 14, no. 1 (2018): 209–36.

98. Compare Eran Magen, Carol S. Dweck, and James J. Gross, "The Hidden Zero Effect: Representing a Single Choice as an Extended Sequence Reduces Impulsive Choice," *Psychological Science* 19, no. 7 (2008): 648–49.

99. Francesca Ducci and David Goldman, "The Genetic Basis of Addictive Disorders," *Psychiatric Clinics* 35, no. 2 (2012): 495–519; Richard C. Crist, Benjamin C. Reiner, and Wade H. Berrettini, "A Review of Opioid Addiction Genetics," *Current Opinion in Psychology* 27 (2019): 31–35; Hedyeh Fazel Tolami et al., "Haplotype-Based Association and in Silico Studies of OPRM1 Gene Variants with Susceptibility to Opioid Dependence Among Addicted Iranians Undergoing Methadone Treatment," *Journal of Molecular Neuroscience* 70 (2020): 504–13; Vadim Yuferov et al., "Search for Genetic Markers and Functional Variants Involved in the Development of Opiate and Cocaine Addiction and Treatment," *Annals of the New York Academy of Sciences* 1187, no. 1 (2010): 184–207; Hang Zhou et al., "Genetic Risk Variants Associated with Comorbid Alcohol Dependence and Major Depression," *JAMA Psychiatry* 74, no. 12 (2017): 1234–41.

100. Brittany M. Priddy et al., "Sex, Strain, and Estrous Cycle Influences on Alcohol Drinking in Rats," *Pharmacology Biochemistry and Behavior* 152 (2017): 61–67; Eleanor B. Towers et al., "Male and Female Mice Develop Escalation of Heroin Intake and Dependence Following Extended Access," *Neuropharmacology* 151 (2019): 189–94; Renata C. N. Marchette et al., "κ-Opioid Receptor Antagonism Reverses Heroin Withdrawal-Induced Hyperalgesia in Male and Female Rats," *Neurobiology of Stress* 14 (2021): 100325.

101. Jeffery S. Mogil, "Sex Differences in Pain and Pain Inhibition: Multiple Explanations of a Controversial Phenomenon," *Nature Reviews*

Neuroscience 13, no. 12 (2012): 859–66; Marilyn E. Carroll and John R. Smethells, "Sex Differences in Behavioral Dyscontrol: Role in Drug Addiction and Novel Treatments," *Frontiers in Psychiatry* 6 (2016): 175.

102. Nancy D. Campbell, "Toward a Critical Neuroscience of 'Addiction,'" *BioSocieties* 5, no. 1 (2010): 89–104; Nancy D. Campbell, *OD: Naloxone and the Politics of Overdose* (MIT Press, 2020).

103. Dana B. Hancock et al., "Human Genetics of Addiction: New Insights and Future Directions," *Current Psychiatry Reports* 20 (2018): 1–17; Orna Levran et al., "A 3'UTR SNP rs885863, a Cis-eQTL for the Circadian Gene VIPR2 and LincRNA 689, Is Associated with Opioid Addiction," *PLoS One* 14, no. 11 (2019): e0224399; Zhou et al., "Genetic Risk Variants Associated with Comorbid Alcohol Dependence."

104. Campbell, "Toward a Critical Neuroscience of 'Addiction.'"

105. Netherland and Hansen, "White Opioids."

106. Sterling and Eyer, "Allostasis"; Schulkin, *Rethinking Homeostasis*; Schulkin, *Adaptation and Well-Being.*

107. David Herzberg, *White Market Drugs: Big Pharma and the Hidden History of Addiction in America* (University of Chicago Press, 2020).

108. Koob and Schulkin, "Addiction and Stress."

109. Joe Herbert, "Peptides in the Limbic System: Neurochemical Codes for Co-Ordinated Adaptive Responses to Behavioural and Physiological Demand," *Progress in Neurobiology* 41, no. 6 (1993): 723–91; Koob and Le Moal, *Neurobiology of Addiction.*

110. Sterling, *What Is Health?*

111. Campbell, "Toward a Critical Neuroscience of 'Addiction'"; Campbell, *OD*; Netherland and Hansen, "White Opioids."

112. Keith Humphreys, *Circles of Recovery: Self-Help Organizations for Addictions* (Cambridge University Press, 2004).

4. SOCIAL HISTORIES AND SOCIAL CONSTRAINTS

1. Kevin Laland, *Darwin's Unfinished Symphony: How Culture Made the Human Mind* (Princeton University Press, 2017); Hazel Rose Markus and Shinobu Kitayama, "Cultures and Selves: A Cycle of Mutual Constitution," *Perspectives on Psychological Science* 5, no. 4 (2010): 420–30;

Simon M. Reader and Kevin N. Laland, "Social Intelligence, Innovation, and Enhanced Brain Size in Primates," *Proceedings of the National Academy of Sciences* 99, no. 7 (2002): 4436–41.

2. Nancy D. Campbell, *OD: Naloxone and the Politics of Overdose* (MIT Press, 2020).

3. Alfred North Whitehead, *The Function of Reason* (Princeton University Press, 1929).

4. Paul Rozin, "The Evolution of Intelligence and Access to the Cognitive Unconscious," in *Progress in Psychobiology and Physiological Psychology*, ed. James M. Sprague and Alan N. Epstein (Academic Press, 1976), 245–80.

5. Randolph M. Nesse and Kent C. Berridge, "Psychoactive Drug Use in Evolutionary Perspective," *Science* 278, no. 5335 (1997): 63–66.

6. Stephen R. Platt, *Imperial Twilight: The Opium War and the End of China's Last Golden Age* (Knopf, 2018).

7. John H. Halpern and David Blistein, *Opium: How an Ancient Flower Shaped and Poisoned Our World* (Hachette, 2019).

8. Mark David Merlin, *On the Trail of the Ancient Opium Poppy* (Associated University Presses, 1984); Aurélie Salavert et al., "Direct Dating Reveals the Early History of Opium Poppy in Western Europe," *Scientific Reports* 10, no. 1 (2020): 20263.

9. Paula Veiga, "Opium: Was It Used as a Recreational Drug in Ancient Egypt?," in *Cultural and Linguistic Transition Explored: Proceedings of the ATrA Closing Workshop*, Trieste, May 25–26, 2016, EUT, 3 (2017): 199–215.

10. Veiga, "Opium."

11. Martin Booth, *Opium: A History* (Macmillan, 1999).

12. Veiga, "Opium."

13. William Shakespeare, *Othello* (Classic Books Company, 2001).

14. Johan W. Goethe, *The Metamorphosis of Plants* (MIT Press, 1790/2009); Alexander Von Humboldt, *Cosmos: A Sketch of the Physical Description of the Universe*, vol. 1, trans. E. C. Otté (Johns Hopkins University Press, 1848/1997); Alexander Von Humboldt and Aimé Bonpland, *Essay on the Geography of Plants* (University of Chicago Press, 2009); Laura Dassow Walls, *The Passage to Cosmos: Alexander von Humboldt and the Shaping of America* (University of Chicago Press, 2019).

15. Erasmus Darwin, *The Botanic Garden* (T. & J. Swords, 1898).

16. Booth, *Opium*.

17. Henry E. Sigerist, "Laudanum in the Works of Paracelsus," *Bulletin of the History of Medicine* 9, no. 5 (1941): 530–44; see also Alethea Hayter, *Opium and the Romantic Imagination* (University of California Press, 1996).

18. William L. White, *Slaying the Dragon: The History of Addiction Treatment and Recovery in America* (Chestnut Health Systems, 1998).

19. W. Ross Albury, "Experiment and Explanation in the Physiology of Bichat and Magendie," *Studies in History of Biology* 1 (1977): 47–131.

20. Claud Bernard, *An Introduction to the Study of Experimental Medicine* (Dover, 1865/1957); see Laurival A. De Luca Jr., "A Critique on the Theory of Homeostasis," *Physiology & Behavior* 247 (2022): 113712.

21. Walter Bradford Cannon, *The Wisdom of the Body* (Norton, 1939); Ernest Henry Starling, "The Croonian Lectures," *Lancet* 26 (1905): 579–83.

22. Curt P. Richter, "Total Self-Regulatory Functions in Animals and Human Beings," *Harvey Lecture Series* 38, no. 63 (1943).

23. Lawrence Coleman Kolb, *Drug Addiction: A Medical Problem* (Charles C. Thomas, 1962); Charles E. Terry, "The Development and Causes of Opium Addiction as a Social Problem," *Journal of Educational Sociology* 4, no. 6 (1931): 335–46.

24. Alonzo Calkins, *Opium and the Opium-Appetite: With Notices of Alcoholic Beverages, Cannabis Indica, Tobacco and Coca, and Tea and Coffee, in Their Hygienic Aspects and Pathologic Relationships* (Lippincott, 1871); see John C. Kramer, "The Opiates: Two Centuries of Scientific Study," *Journal of Psychedelic Drugs* 12, no. 2 (1980): 89–103.

25. Campbell, *OD*.

26. Bernard Lazarus, "A Contribution to the Therapeutic Action of Heroin," *Boston Medical and Surgical Journal* 143, no. 24 (1900): 600–602.

27. Peter Sterling and Joseph Eyer, "Allostasis: A New Paradigm to Explain Arousal Pathology," in *Handbook of Life Stress, Cognition and Health*, ed. Shirley Fisher and James Reason (Wiley, 1988), 629–49; Jay Schulkin, *Rethinking Homeostasis* (MIT Press, 2003).

28. Samuel Hopkins Adams, *The Great American Fraud* (American Medical Association, 1912); see David Herzberg, *White Market Drugs: Big Pharma and the Hidden History of Addiction in America* (University of Chicago Press, 2020).

29. Neil Vickers, *Coleridge and the Doctors: 1795–1806* (Oxford University Press, 2004).

30. Barry Milligan, "Morphine-Addicted Doctors, the English Opium-Eater, and Embattled Medical Authority," *Victorian Literature and Culture* 33, no. 2 (2005): 541–53.

31. Virginia Berridge and Griffith Edwards, *Opium and the People: Opiate Use in Nineteenth-Century England* (Free Association Books, 1981); Virginia Berridge and Sarah Mars, "History of Addictions," *Journal of Epidemiology & Community Health* 58, no. 9 (2004): 747–50.

32. David T. Courtwright, *Forces of Habit: Drugs and the Making of the Modern World* (Harvard University Press, 2001).

33. Charles Stewart Roberts, "H. L. Mencken and the Four Doctors: Osler, Halsted, Welch, and Kelly," *Baylor University Medical Center Proceedings* 23, no. 4 (2010): 377–88.

34. William Osler, *The Principles and Practice of Medicine* (Appleton, 1892).

35. Booth, *Opium*, 81; Michael Bliss, *William Osler: A Life in Medicine* (Oxford University Press, 1999).

36. See David T. Courtwright, *Dark Paradise: A History of Opiate Addiction in America* (Harvard University Press, 1982); Courtwright, *Forces of Habit*.

37. Paul L. Schiff, "Opium and Its Alkaloids," *American Journal of Pharmaceutical Education* 66, no. 2 (2002): 188–96.

38. Caroline Jean Acker, *Creating the American Junkie: Addiction Research in the Classic Era of Narcotic Control* (Johns Hopkins University Press, 2003).

39. Herzberg, *White Market Drugs*, 89.

40. Berridge and Edwards, *Opium and the People*; Berridge and Mars, "History of Addictions."

41. John Masson Gulland and Robert Robinson, "Constitution of Codeine and Thebaine," *Memoirs and Proceedings of the Manchester Literary and Philosophical Society* 69 (1925): 79–86; Gavril W. Pasternak and Ying-Xian Pan, "Mu Opioids and Their Receptors: Evolution of a Concept," *Pharmacological Reviews* 65, no. 4 (2013): 1257–1317.

42. See Pasternak and Pan, "Mu Opioids and Their Receptors."

43. Herzberg, *White Market Drugs*.

44. Lukasz Kamienski, *Shooting Up: A Short History of Drugs and War* (Oxford University Press, 2016).

45. Jay Schulkin, *Oliver Wendell Holmes Jr., Pragmatism and Neuroscience* (Palgrave MacMillan, 2019).
46. Santiago Ramon Y. Cajal, *Recollections of My Life*, vol. 8 (MIT Press, 1989), 285.
47. Herzberg, *White Market Drugs*.
48. Jonathan S. Jones, "Opium Slavery," *Journal of the Civil War Era* 10, no. 2 (2020): 185–212.
49. David T. Courtwright, "The Hidden Epidemic: Opiate Addiction and Cocaine Use in the South, 1860–1920," *Journal of Southern History* 49, no. 1 (1983): 57–72.
50. Mary Boykin Chesnut, Comer Vann Woodward, and Elisabeth Muhlenfeld, *The Private Mary Chesnut: The Unpublished Civil War Diaries*, vol. 773 (Oxford University Press, 1984), 95.
51. Herzberg, *White Market Drugs*.
52. Herzberg, *White Market Drugs*; Jones, "Opium Slavery."
53. Courtwright, *Dark Paradise*; Courtwright, *Forces of Habit*; John C. Burnham, *Bad Habits: Drinking, Smoking, Taking Drugs, Gambling, Sexual Misbehavior, and Swearing in American History* (New York University Press, 1993).
54. Oscar Jennings, *On the Cure of the Morphine Habit* (Balliere, Tindall and Cox, 1890); Harry Hubbell Kane, *The Hypdermic Injection of Morphine* (Chas L. Bermingham, 1880), 31.
55. Booth, *Opium*.
56. Berridge and Edwards, *Opium and the People*; Berridge and Mars, "History of Addictions."
57. David Sarokin and Jay Schulkin, *The Corporation: Its History and Future* (Cambridge Scholars Publishing, 2020).
58. Courtwright, *Forces of Habit*.
59. Thomas Dormandy, *Opium: Reality's Dark Dream* (Yale University Press, 2012).
60. Courtwright, *Forces of Habit*, 35.
61. Courtwright, *Dark Paradise*; Courtwright, *Forces of Habit*; Herzberg, *White Market Drugs*.
62. See David F. Musto, *The American Disease: Origins of Narcotic Control* (Oxford University Press, 1987).
63. Courtwright, *Dark Paradise*; Courtwright, *Forces of Habit*; Campbell, *OD*.

64. Acker, *Creating the American Junkie.*

65. Herzberg, *White Market Drugs.*

66. Dormandy, *Opium.*

67. Thomas de Quincey, *Confessions of an English Opium Eater* (Penguin, 1821/1994); see Frances Wilson, *Guilty Thing: The Life of Thomas De Quincey* (Farrar, Straus and Giroux, 2016).

68. Hayter, *Opium and the Romantic Imagination.*

69. Wilson, *Guilty Thing,* 114.

70. Courtwright, *Dark Paradise.*

71. Hayter, *Opium and the Romantic Imagination,* 166.

72. Joseph Brent, *Charles Sanders Peirce: A Life* (Indiana University Press, 1998).

73. Courtwright, *Dark Paradise;* Courtwright, *Forces of Habit;* David T. Courtwright, *The Age of Addiction: How Bad Habits Became Big Business* (Harvard University Press, 2019).

74. David Herzberg, "Entitled to Addiction? Pharmaceuticals, Race, and America's First Drug War," *Bulletin of the History of Medicine* 91, no. 3 (2017): 586–623; Jones, "Opium Slavery."

75. The Velvet Underground, "Heroin," track 1, side 2, *The Velvet Underground and Nico,* Verve Records, 1967, LP.

76. The Beatles, "Happiness Is a Warm Gun," track 8, side 1, *The Beatles,* EMI, 1968, LP.

77. Campbell, *OD;* Herzberg, *White Market Drugs.*

78. Courtwright, *The Age of Addiction.*

79. Merianne R. Spencer, Arialdi M. Miniño, and Margaret Warner, "Drug Overdose Deaths in the United States, 2001–2021," *NCHS Data Brief,* no. 457 (2022).

80. Courtwright, *Dark Paradise;* Courtwright, *The Age of Addiction;* David T. Courtwright, Herman Joseph, and Don Des Jarlais, *Addicts Who Survived: An Oral History of Narcotic Use in America Before 1965* (University of Tennessee Press, 1989).

81. Courtwright, *Dark Paradise;* Courtwright, *Forces of Habit.*

5. MANAGING PAIN

1. Travis N. Rieder, *In Pain: A Bioethicist's Personal Struggle with Opioids* (Harper Collins, 2019), 9.

2. Rieder, *In Pain*, 34.

3. Rieder, *In Pain*.

4. David Herzberg, *White Market Drugs: Big Pharma and the Hidden History of Addiction in America* (University of Chicago Press, 2020).

5. Jonathan S. Jones, "Opium Slavery," *Journal of the Civil War Era* 10, no. 2 (2020): 185–212.

6. Jeremy A. Greene and David Herzberg, "Hidden in Plain Sight: Marketing Prescription Drugs to Consumers in the Twentieth Century," *American Journal of Public Health* 100, no. 5 (2010): 793–803; Nancy D. Campbell, *OD: Naloxone and the Politics of Overdose* (MIT Press, 2020).

7. Sarah DeWeerdt, "Tracing the US Opioid Crisis to Its Roots," *Nature* 573, no. 7773 (2019): S10–S12.

8. Jason Dana and George Loewenstein, "A Social Science Perspective on Gifts to Physicians from Industry," *Jama* 290, no. 2 (2003): 252–55.

9. Britta L. Anderson et al., "Factors Associated with Physicians' Reliance on Pharmaceutical Sales Representatives," *Academic Medicine* 84, no. 8 (2009): 994–1002; Gabriel K. Silverman et al., "Failure to Discount for Conflict of Interest When Evaluating Medical Literature: A Randomised Trial of Physicians," *Journal of Medical Ethics* 36, no. 5 (2010): 265–70.

10. Dana and Loewenstein, "A Social Science Perspective on Gifts to Physicians from Industry"; Herzberg, *White Market Drugs*.

11. Ronald T. Libby, *Treating Doctors as Drug Dealers: The DEA's War on Prescription Painkillers* (Policy Analysis, 2005), 545.

12. George B. Richardson, Taheera N. Blount, and Blair S. Hanson-Cook, "Life History Theory and Recovery from Substance Use Disorder," *Review of General Psychology* 23, no. 2 (2019): 263–74.

13. Ajay Manhapra and William C. Becker, "Pain and Addiction: An Integrative Therapeutic Approach," *Medical Clinics* 102, no. 4 (2018): 745–63.

14. Kathleen M. Foley, "The Treatment of Cancer Pain," *New England Journal of Medicine* 313, no. 2 (1985): 84–95.

15. Deborah Dowell, Tamara M. Haegerich, and Roger Chou, "CDC Guideline for Prescribing Opioids for Chronic Pain—United States, 2016," *Jama* 315, no. 15 (2016): 1624–45.

16. Salimah H. Meghani and Neha Vapiwala, "Bridging the Critical Divide in Pain Management Guidelines from the CDC, NCCN, and ASCO for Cancer Survivors," *JAMA Oncology* 4, no. 10 (2018): 1323–24.

17. Herzberg, *White Market Drugs*.

18. Anna Lembke, *Drug Dealer, MD: How Doctors were Duped, Patients got Hooked, and Why It's so Hard to Stop* (Johns Hopkins University Press, 2016).

19. Lembke, *Drug Dealer, MD*.

20. David Sarokin and Jay Schulkin, *The Corporation: Its History and Future* (Cambridge Scholars Publishing, 2020).

21. Travis N. Rieder, "Pain Medicine During an Opioid Epidemic Needs More Transparency, not Less," *AJOB Neuroscience* 9, no. 3 (2018): 183–206.

22. Lembke, *Drug Dealer, MD*.

23. Sarokin and Schulkin, *The Corporation*.

24. George Loewenstein, *Exotic Preferences: Behavioral Economics and Human Motivation* (Oxford University Press, 2007).

25. Campbell, *OD*; David T. Courtwright, *Dark Paradise: A History of Opiate Addiction in America* (Harvard University Press, 1982); David T. Courtwright, *Forces of Habit: Drugs and the Making of the Modern World* (Harvard University Press, 2001).

26. Kelly M. Hoffman et al., "Racial Bias in Pain Assessment and Treatment Recommendations, and False Beliefs About Biological Differences Between Blacks and Whites," *Proceedings of the National Academy of Sciences* 113, no. 16 (2016): 4296–4301; Nancy E. Morden et al., "Racial Inequality in Prescription Opioid Receipt—Role of Individual Health Systems," *New England Journal of Medicine* 385, no. 4 (2021): 342–51.

27. Sarokin and Schulkin, *The Corporation*.

28. Andrew K. Chang et al., "Effect of a Single Dose of Oral Opioid and Nonopioid Analgesics on Acute Extremity Pain in the Emergency Department: A Randomized Clinical Trial," *Jama* 318, no. 17 (2017): 1661–67; Silva Minozzi, Laura Amato, and Marina Davoli, "Development of Dependence Following Treatment with Opioid Analgesics for Pain Relief: A Systematic Review," *Addiction* 108, no. 4 (2013): 688–98.

29. Compare Jane Porter and Hershel Jick, "Addiction Rare in Patients Treated with Narcotics," *New England Journal of Medicine* 302, no. 2 (1980): 123.

30. Vilayanur S. Ramachandran and William Hirstein, "The Perception of Phantom Limbs: The DO Hebb Lecture," *Brain: A Journal of Neurology* 121, no. 9 (1998): 1603–30.

31. Jonathan D. Moreno, *Undue Risk: Secret State Experiments on Humans* (Freeman, 2000).

32. Charles Darwin, *The Expression of the Emotions in Man and Animals* (Oxford University Press, 1872/1998).

33. Hoffman et al., "Racial Bias in Pain Assessment and Treatment Recommendations."

34. Bryce Huebner, "The Emptiness of Anger," *Journal of Buddhist Philosophy*, no. 3 (2021): 50–67; Adam Smith, *The Theory of Moral Sentiments* (Liberty Classics, 1759/1982).

35. Smith, *The Theory of Moral Sentiments.*

36. Predrag Petrovic et al., "Placebo and Opioid Analgesia—Imaging a Shared Neuronal Network," *Science* 295, no. 5560 (2002): 1737–40.

37. Sophie Rosenkjær et al., "Expectations: How and When Do They Contribute to Placebo Analgesia?," *Frontiers in Psychiatry* 13 (2022): 817179; Irene Tracey and Patrick W. Mantyh, "The Cerebral Signature for Pain Perception and Its Modulation," *Neuron* 55, no. 3 (2007): 377–91.

38. Jon-Kar Zubieta et al., "Placebo Effects Mediated by Endogenous Opioid Activity on μ-Opioid Receptors," *Journal of Neuroscience* 25, no. 34 (2005): 7754–62.

39. Randolph M. Nesse and Jay Schulkin, "An Evolutionary Medicine Perspective on Pain and Its Disorders," *Philosophical Transactions of the Royal Society B* 374, no. 1785 (2019): 20190288.

40. Luana Colloca et al., "Placebo Analgesia: Psychological and Neurobiological Mechanisms," *Pain* 154, no. 4 (2013): 511–14.

41. Elaine M. Jennings et al., "Stress-Induced Hyperalgesia," *Progress in Neurobiology* 121 (2014): 1–18.

42. A. Vania Apkarian, Marwan N. Baliki, and Paul Y. Geha, "Towards a Theory of Chronic Pain," *Progress in Neurobiology* 87, no. 2 (2009): 81–97; Michael Costigan, Joachim Scholz, and Clifford J. Woolf, "Neuropathic Pain: A Maladaptive Response of the Nervous System to Damage," *Annual Review of Neuroscience* 32, no. 1 (2009): 1–32.

43. Jelle Zorn et al., "Cognitive Defusion Is a Core Cognitive Mechanism for the Sensory-Affective Uncoupling of Pain During Mindfulness Meditation," *Psychosomatic Medicine* 83, no. 6 (2021): 566–78.

44. Simon B. Goldberg et al., "Prevalence of Meditation-Related Adverse Effects in a Population-Based Sample in the United States," *Psychotherapy Research* 32, no. 3 (2022): 291–305.

45. Aviel Goodman, "Neurobiology of Addiction: An Integrative Review," *Biochemical Pharmacology* 75, no. 1 (2008): 266–322.

46. Nancy D. Campbell, "Toward a Critical Neuroscience of 'Addiction,'" *BioSocieties* 5, no. 1 (2010): 89–104.

6. ACCOUNTABILITY AND REDUCING HARM

1. Roy Porter and Mikulas Teich, eds., *Drugs and Narcotics in History* (Cambridge University Press, 1995).

2. Claire D. Clark, *The Recovery Revolution: The Battle over Addiction Treatment in the United States* (Columbia University Press, 2017), 211.

3. Caroline Jean Acker, *Creating the American Junkie: Addiction Research in the Classic Era of Narcotic Control* (Johns Hopkins University Press, 2003).

4. Nancy D. Campbell, J. P. Olsen, and Luke Walden, *The Narcotic Farm: The Rise and Fall of America's First Prison for Drug Addicts* (University Press of Kentucky, 2021).

5. Compare Vincent P. Dole and Marie E. Nyswander, "Heroin Addiction—A Metabolic Disease," *Archives of Internal Medicine* 120, no. 1 (1967): 19–24; Abraham Wikler, "On the Nature of Addiction and Habituation," *British Journal of Addiction to Alcohol & Other Drugs* 57, no. 2 (1961): 73–79.

6. Nancy D. Campbell, "Toward a Critical Neuroscience of 'Addiction,'" *BioSocieties* 5, no. 1 (2010): 94.

7. Campbell, Olsen, and Walden, *The Narcotic Farm*.

8. Nancy D. Campbell, *OD: Naloxone and the Politics of Overdose* (MIT Press, 2020); Nancy D. Campbell and Anne M. Lovell, "The History of the Development of Buprenorphine as an Addiction Therapeutic," *Annals of the New York Academy of Sciences* 1248, no. 1 (2012): 124–39.

9. Campbell, Olsen, and Walden, *The Narcotic Farm*; Jonathan D. Moreno, *Undue Risk: Secret State Experiments on Humans* (Freeman, 2000).

10. Peter Andreas, "Drugs and War: What Is the Relationship?," *Annual Review of Political Science* 22, no. 1 (2019): 57–73.

11. David T. Courtwright, *The Age of Addiction: How Bad Habits Became Big Business* (Harvard University Press, 2019), 118.

12. Jay Schulkin, "A Pragmatist Perspective on Brains, Trust, and Choice," *Journal of Speculative Philosophy* 37, no. 1 (2023): 61–80.

13. Cullen L. Schmid et al., "Bias Factor and Therapeutic Window Correlate to Predict Safer Opioid Analgesics," *Cell* 171, no. 5 (2017): 1165–75.

14. Caitlin E. Martin, Mishka Terplan, and Elizabeth E. Krans, "Pain, Opioids, and Pregnancy: Historical Context and Medical Management," *Clinics in Perinatology* 46, no. 4 (2019): 833–47.

15. Nancy D. Campbell, "The Construction of Pregnant Drug-Using Women as Criminal Perpetrators," *Fordham Urban Law Journal* 33 (2005): 463; Nancy D. Campbell, "When Should Screening and Surveillance Be Used During Pregnancy?," *AMA Journal of Ethics* 20, no. 3 (2018): 288–95; Nancy D. Campbell and Elizabeth Ettorre, *Gendering Addiction: The Politics of Drug Treatment in a Neurochemical World* (Springer, 2011).

16. Campbell and Ettorre, *Gendering Addiction*.

17. Shalini Arunogiri et al., "Managing Opioid Dependence in Pregnancy: A General Practice Perspective," *Australian Family Physician* 42, no. 10 (2013): 713–16; Martin, Terplan, and Krans, "Pain, Opioids, and Pregnancy."

18. Hendrée E. Jones et al., "Buprenorphine Treatment of Opioid-Dependent Pregnant Women: A Comprehensive Review," *Addiction* 107 (2012): 5–27.

19. Kristen L. Benninger, Jennifer M. McAllister, and Stephanie L. Merhar, "Neonatal Opioid Withdrawal Syndrome: An Update on Developmental Outcomes," *Clinics in Perinatology* 50, no. 1 (2023): 17–29.

20. Benninger, McAllister, and Merhar, "Neonatal Opioid Withdrawal Syndrome."

21. Benninger, McAllister, and Merhar, "Neonatal Opioid Withdrawal Syndrome"; compare Sarah S. Richardson, *The Maternal Imprint: The Contested Science of Maternal-Fetal Effects* (University of Chicago Press, 2021).

22. Joe Dooley et al., "Buprenorphine-Naloxone Use in Pregnancy for Treatment of Opioid Dependence: Retrospective Cohort Study of 30 Patients," *Canadian Family Physician* 62, no. 4 (2016): e194–e200.

23. Maia Szalavitz, *Undoing Drugs: The Untold Story of Harm Reduction and the Future of Addiction* (Hachette, 2021), 219.

24. Campbell, *OD*; Kathryn F. Hawk, Federico E. Vaca, and Gail D'Onofrio, "Focus: Addiction: Reducing Fatal Opioid Overdose: Prevention, Treatment and Harm Reduction Strategies," *Yale Journal of Biology and Medicine* 88, no. 3 (2015): 235–45.

25. John H. Halpern and David Blistein, *Opium: How an Ancient Flower Shaped and Poisoned Our World* (Hachette, 2019); Keith Humphreys, *Circles of Recovery: Self-Help Organizations for Addictions* (Cambridge University Press, 2004); William L. White, *Slaying the Dragon: The History of Addiction Treatment and Recovery in America* (Chestnut Health Systems, 1998).

26. Nora D. Volkow, George F. Koob, and A. Thomas McLellan, "Neurobiologic Advances from the Brain Disease Model of Addiction," *New England Journal of Medicine* 374, no. 4 (2016): 363–71.

27. Bryce Huebner and Jay Schulkin, *Biological Cognition* (Cambridge University Press, 2022); Jonathan D. Moreno and Jay Schulkin, *The Brain in Context: A Pragmatic Guide to Neuroscience* (Columbia University Press, 2020).

28. Nancy D. Campbell, "The Impact of Changes in Neuroscience and Research Ethics on the Intellectual History of Addiction Research," in *Addiction Neuroethics* (Academic Press, 2012), 197–213; Hazel Rose Markus and Shinobu Kitayama, "Cultures and Selves: A Cycle of Mutual Constitution," *Perspectives on Psychological Science* 5, no. 4 (2010): 420–30.

29. Roger B. Fillingim et al., "Sex, Gender, and Pain: A Review of Recent Clinical and Experimental Findings," *Journal of Pain* 10, no. 5 (2009): 447–85; Jeffrey S. Mogil and Benjamin Kest, "Sex Differences in Opioid Analgesia: Of Mice and Women," *Pain Forum* 8, no. 1 (1999): 48–50.

30. Lester Darryl Geneviève et al., "Structural Racism in Precision Medicine: Leaving No One Behind," *BMC Medical Ethics* 21 (2020): 1–13.

31. Szalavitz, *Undoing Drugs*, 238.

32. Szalavitz. *Undoing Drugs*, 272.

33. Anne M. Fletcher, *Inside Rehab: The Surprising Truth About Addiction Treatment—And How to Get Help That Works* (Penguin, 2013).

34. Campbell, *OD*.

35. Campbell, *OD*, 73.

36. Michelle Alexander, *The New Jim Crow: Mass Incarceration in the Age of Colorblindness* (New Press, 2020); Ellen A. Donnelly et al., "Revisiting Neighborhood Context and Racial Disparities in Drug Arrests Under the Opioid Epidemic," *Race and Justice* 12, no. 2 (2022): 322–43; Shytierra Gaston, "Enforcing Race: A Neighborhood-Level Explanation of Black–White Differences in Drug Arrests," *Crime & Delinquency* 65, no. 4 (2019): 499–526.

37. Campbell, *OD*; Szalavitz, *Undoing Drugs*.

38. Szalavitz, *Undoing Drugs*.

39. David T. Courtwright, *Dark Paradise: A History of Opiate Addiction in America* (Harvard University Press, 1982), 2.

40. Szalavitz, *Undoing Drugs*.

41. Alexander, *The New Jim Crow*.

42. Szalavitz, *Undoing Drugs*, 307.

43. Huebner and Schulkin, *Biological Cognition*.

44. Joan B. Silk, "The Adaptive Value of Sociality in Mammalian Groups," *Philosophical Transactions of the Royal Society B: Biological Sciences* 362, no. 1480 (2007): 539–59.

45. Huebner and Schulkin, *Biological Cognition*; Nina Marsh et al., "Oxytocin and the Neurobiology of Prosocial Behavior," *Neuroscientist* 27, no. 6 (2021): 604–19.

46. Mark Dingemanse et al., "Beyond Single-Mindedness: A Figure-Ground Reversal for the Cognitive Sciences," *Cognitive Science* 47, no. 1 (2023): e13230.

47. Maurice Hamington, *Embodied Care: Jane Addams, Maurice Merleau-Ponty, and Feminist Ethics* (University of Illinois Press, 2004).

48. Ayna B. Johansen et al., "Practical Support Aids Addiction Recovery: The Positive Identity Model of Change," *BMC Psychiatry* 13 (2013): 1–11.

49. Laura M. Harvey et al., "Psychosocial Intervention Utilization and Substance Abuse Treatment Outcomes in a Multisite Sample of Individuals Who Use Opioids," *Journal of Substance Abuse Treatment* 112 (2020): 68–75.

50. Bryce Huebner, "Planning and Prefigurative Practice," in *The Philosophy of Daniel Dennett* (Oxford University Press, 2017), 295–327; Peter Kropotkin, *Mutual Aid: A Factor of Evolution* (Black Rose, 1904/2021).

51. Jonathan D. Moreno, *Deciding Together: Bioethics and Moral Consensus* (Oxford University Press, 1995).

52. Huebner, "Planning and Prefigurative Practice."

53. Fan Wang et al., "A Systematic Review and Meta-Analysis of 90 Cohort Studies of Social Isolation, Loneliness and Mortality," *Nature Human Behaviour* 7, no. 8 (2023): 1307–19.

54. Robert Fiorentine and Maureen P. Hillhouse, "Drug Treatment and 12-Step Program Participation: The Additive Effects of Integrated Recovery Activities," *Journal of Substance Abuse Treatment* 18, no. 1

(2000): 65–74; Keith Humphreys et al., "Impact of 12 Step Mutual Help Groups on Drug Use Disorder Patients Across Six Clinical Trials," *Drug and Alcohol Dependence* 215 (2020): 108213.

55. George F. Koob, "Drug Addiction: Hyperkatifeia/Negative Reinforcement as a Framework for Medications Development," *Pharmacological Reviews* 73, no. 1 (2021): 163–201.

56. Emile Durkheim, *Suicide: A Study in Sociology*, trans. J. A. Spaulding and G. Simpson (Free Press, 1897/1951).

57. Jeffrey Foote et al., *Beyond Addiction: How Science and Kindness Help People Change* (Simon & Schuster, 2014).

58. Bruce S. McEwen, "In Pursuit of Resilience: Stress, Epigenetics, and Brain Plasticity," *Annals of the New York Academy of Sciences* 1373, no. 1 (2016): 56–64; Bruce S. McEwen, "Allostasis and the Epigenetics of Brain and Body Health over the Life Course: The Brain on Stress," *JAMA Psychiatry* 74, no. 6 (2017): 551–52.

CONCLUSION: PURSUING FREEDOM

1. Richard Holton and Kent Berridge, "Compulsion and Choice in Addiction," *Addiction and Choice: Rethinking the Relationship* (Oxford University Press, 2017), 153–70; George F. Koob, "Drug Addiction: Hyperkatifeia/Negative Reinforcement as a Framework for Medications Development," *Pharmacological Reviews* 73, no. 1 (2021): 163–201.

2. Bryce Huebner, "Planning and Prefigurative Practice," in *The Philosophy of Daniel Dennett* (Oxford University Press, 2017), 295–327; Jay Schulkin, *Effort: A Behavioral Neuroscience Perspective on the Will* (Erlbaum, 2007).

3. Daniel M. Wegner, *The Illusion of Conscious Will* (MIT Press, 2002); Robert M. Sapolsky, *Behave* (Penguin, 2017).

4. Daniel C. Dennett, *Freedom Evolves* (Penguin UK, 2004).

5. Jean Paul Sartre, *Being and Nothingness* (Philosophical Library, 1956).

6. Holton and Berridge, "Compulsion and Choice in Addiction," 155.

7. Schulkin, *Effort*.

8. Gerd Gigerenzer, *Adaptive Thinking* (Oxford University Press, 2000); Daniel Kahneman, *Thinking, Fast and Slow* (Farrar, Straus and Giroux, 2011); George Loewenstein, *Exotic Preferences: Behavioral Economics and Human Motivation* (Oxford University Press, 2007).

9. Gary J. Badger et al., "Altered States: The Impact of Immediate Craving on the Valuation of Current and Future Opioids," *Journal of Health Economics* 26, no. 5 (2007): 865–76.

10. Terry E. Robinson and Kent C. Berridge, "The Neural Basis of Drug Craving: An Incentive-Sensitization Theory of Addiction," *Brain Research Reviews* 18, no. 3 (1993): 247–91.

11. Holton and Berridge, "Compulsion and Choice in Addiction."

12. Brendan Dill and Richard Holton, "The Addict in Us All," *Frontiers in Psychiatry* 5 (2014): 139.

13. Benedict de Spinoza, "Ethics," in *Spinoza: Complete Works*, trans. Samuel Shirley, ed. Michael Morgan (Hackett, 1677/2002), 239.

14. Spinoza, "Ethics," 320.

15. Karin Meyers, "Free Persons, Empty Selves: Freedom and Agency in Light of the Two Truths," in *Free Will, Agency, and Selfhood in Indian Philosophy*, ed. Matthew R. Dasti and Edwin F. Bryant (Oxford University Press, 2013), 41–67.

16. Bryce Huebner, "The Emptiness of Anger," *Journal of Buddhist Philosophy*, no. 3 (2021): 50–67; Emily McRae, "Emotions and Choice: Lessons from Tsongkhapa," in *Buddhist Perspectives on Free Will: Agentless Agency?*, ed. Rick Repetti (Routledge, 2016), 170–81.

17. Huebner, "Planning and Prefigurative Practice"; Mark L. Johnson and Jay Schulkin, *Mind in Nature: John Dewey, Cognitive Science, and a Naturalistic Philosophy for Living* (MIT Press, 2023).

18. George Ainslie, *Breakdown of Will* (Cambridge University Press, 2001).

19. Dennett, *Freedom Evolves*, 207.

20. Marieke A. Adriaanse et al., "Breaking Habits with Implementation Intentions: A Test of Underlying Processes," *Personality and Social Psychology Bulletin* 37, no. 4 (2011): 502–13; Molly J. Crockett et al., "Restricting Temptations: Neural Mechanisms of Precommitment," *Neuron* 79, no. 2 (2013): 391–401; Peter M. Gollwitzer, "Implementation Intentions: Strong Effects of Simple Plans," *American Psychologist* 54, no. 7 (1999): 493.

21. Sam J. Gilbert et al., "Separable Brain Systems Supporting Cued Versus Self-Initiated Realization of Delayed Intentions," *Journal of Experimental Psychology: Learning, Memory, and Cognition* 35, no. 4 (2009): 905.

22. Peter M. Gollwitzer, "Mindset Theory of Action Phases," *Handbook of Theories of Social Psychology* 1 (2012): 541.

23. Jonathan D. Moreno, *Impromptu Man: J. L. Moreno and the Origins of Psychodrama, Encounter Culture, and the Social Network* (Bellevue Literary Press, 2014).

24. Bronwyn Tarr et al., "Music and Social Bonding: 'Self-Other' Merging and Neurohormonal Mechanisms," *Frontiers in Psychology* 5 (2014): 1096.

25. Nancy D. Campbell, "Toward a Critical Neuroscience of 'Addiction,'" *BioSocieties* 5, no. 1 (2010): 89–104; Thích Nhất Hạnh, *Understanding Our Mind: 50 Verses on Buddhist Psychology* (Parallax, 2001).

26. Gene M. Heyman, *Addiction: A Disorder of Choice* (Harvard University Press, 2009).

27. Hazel Rose Markus and Shinobu Kitayama, "Cultures and Selves: A Cycle of Mutual Constitution," *Perspectives on Psychological Science* 5, no. 4 (2010): 420–30.

28. Dennett, *Freedom Evolves*; Huebner, "Planning and Prefigurative Practice."

29. David T. Courtwright, *The Age of Addiction: How Bad Habits Became Big Business* (Harvard University Press, 2019).

30. David Herzberg, *White Market Drugs: Big Pharma and the Hidden History of Addiction in America* (University of Chicago Press, 2020), 9.

31. Maia Szalavitz, *Undoing Drugs: The Untold Story of Harm Reduction and the Future of Addiction* (Hachette, 2021).

32. T. H. Mason and W. B. Hamby, "Relief of Morphine Addiction by Prefrontal Lobotomy," *Journal of the American Medical Association* 136, no. 16 (1948): 1039–40.

33. Barry J. Everitt and Trevor W. Robbins, "Drug Addiction: Updating Actions to Habits to Compulsions Ten Years On," *Annual Review of Psychology* 67, no. 1 (2016): 23–50.

34. Markus Heilig et al., "Addiction as a Brain Disease Revised: Why It Still Matters, and the Need for Consilience," *Neuropsychopharmacology* 46, no. 10 (2021): 1715–23.

35. Sally Satel and Scott Lilienfeld, *Brainwashed: The Seductive Appeal of Mindless Neuroscience* (Basic Books, 2013).

36. Carl Erik Fisher, *The Urge: Our History of Addiction* (Penguin, 2022).

REFERENCES

Acker, Caroline Jean. *Creating the American Junkie: Addiction Research in the Classic Era of Narcotic Control.* Johns Hopkins University Press, 2003.

Adams, Samuel Hopkins. *The Great American Fraud.* American Medical Association, 1912.

Adriaanse, Marieke A., Peter M. Gollwitzer, Denise T. D. De Ridder et al. "Breaking Habits with Implementation Intentions: A Test of Underlying Processes." *Personality and Social Psychology Bulletin* 37, no. 4 (2011): 502–13.

Ainslie, George. *Breakdown of Will.* Cambridge University Press, 2001.

Ainslie, George, and John Monterosso. "A Marketplace in the Brain?" *Science* 306, no. 5695 (2004): 421–23.

Albury, W. Ross. "Experiment and Explanation in the Physiology of Bichat and Magendie." *Studies in History of Biology* 1 (1977): 47–131.

Alexander, Michelle. *The New Jim Crow: Mass Incarceration in the Age of Colorblindness.* New Press, 2020.

Ambron, Richard. *The Brain and Pain: Breakthroughs in Neuroscience.* Columbia University Press, 2022.

Anderson, Britta L., Gabriel K. Silverman, George F. Loewenstein et al. "Factors Associated with Physicians' Reliance on Pharmaceutical Sales Representatives." *Academic Medicine* 84, no. 8 (2009): 994–1002.

Andreas, Peter. "Drugs and War: What Is the Relationship?" *Annual Review of Political Science* 22, no. 1 (2019): 57–73.

Apkarian, A. Vania, Marwan N. Baliki, and Paul Y. Geha. "Towards a Theory of Chronic Pain." *Progress in Neurobiology* 87, no. 2 (2009): 81–97.

Arunogiri, Shalini, Lea Foo, Matthew Frei, and Dan I. Lubman. "Managing Opioid Dependence in Pregnancy: A General Practice Perspective." *Australian Family Physician* 42, no. 10 (2013): 713–16.

Asok, Arun, Adam Draper, Alexander F. Hoffman et al. "Optogenetic Silencing of a Corticotropin-Releasing Factor Pathway from the Central Amygdala to the Bed Nucleus of the Stria Terminalis Disrupts Sustained Fear." *Molecular Psychiatry* 23, no. 4 (2018): 914–22.

Atran, Scott, and Douglas Medin. *The Native Mind and the Cultural Construction of Nature.* MIT Press, 2008.

Badger, Gary J., Warren K. Bickel, Louis A. Giordano et al. "Altered States: The Impact of Immediate Craving on the Valuation of Current and Future Opioids." *Journal of Health Economics* 26, no. 5 (2007): 865–76.

Baumgartner, Hannah M., Madeliene Granillo, Jay Schulkin, and Kent C. Berridge. "Corticotropin Releasing Factor (CRF) Systems: Promoting Cocaine Pursuit Without Distress via Incentive Motivation." *PLoS One* 17, no. 5 (2022): e0267345.

Baumgartner, Hannah M., Jay Schulkin, and Kent C. Berridge. "Activating Corticotropin-Releasing Factor Systems in the Nucleus Accumbens, Amygdala, and Bed Nucleus of Stria Terminalis: Incentive Motivation or Aversive Motivation?" *Biological Psychiatry* 89, no. 12 (2021): 1162–75.

Bechara, Antoine. "Decision Making, Impulse Control and Loss of Willpower to Resist Drugs: A Neurocognitive Perspective." *Nature Neuroscience* 8, no. 11 (2005): 1458–63. https://doi.org/10.1038/NN1584.

Bechara, Antoine. "The Role of Emotion in Decision-Making: Evidence from Neurological Patients with Orbitofrontal Damage." *Brain and Cognition* 55, no. 1 (2004): 30–40.

Beck, Aaron T. *Depression: Clinical, Experimental and Theoretical Aspects.* Harper & Row, 1967.

Benninger, Kristen L., Jennifer M. McAllister, and Stephanie L. Merhar. "Neonatal Opioid Withdrawal Syndrome: An Update on Developmental Outcomes." *Clinics in Perinatology* 50, no. 1 (2023): 17–29.

Bernard, Claud. *An Introduction to the Study of Experimental Medicine.* Dover, 1865/1957.

Berns, Gregory S., David Laibson, and George Loewenstein. "Intertemporal Choice–Toward an Integrative Framework." *Trends in Cognitive Sciences* 11, no. 11 (2007): 482–88.

Berridge, Kent C. "Affective Valence in the Brain: Modules or Modes?" *Nature Reviews Neuroscience* 20, no. 4 (2019): 225–34. https://doi.org/10.1038/S41583 -019-0122-8.

Berridge, Kent C. "From Prediction Error to Incentive Salience: Mesolimbic Computation of Reward Motivation." *European Journal of Neuroscience* 35, no. 7 (2012): 1124–43.

Berridge, Kent C. "Is Addiction a Brain Disease?" *Neuroethics* 10 (2017): 29–33. https://doi.org/10.1007/S12152-016-9286-3.

Berridge, Kent C. "Is Addiction a Brain Disease? The Incentive-Sensitization View." In *Evaluating the Brain Disease Model of Addiction*, 74–86. Routledge, 2022.

Berridge, Kent C., Francis W. Flynn, Jay Schulkin, and Harvey J. Grill. "Sodium Depletion Enhances Salt Palatability in Rats." *Behavioral Neuroscience* 98, no. 4 (1984): 652.

Berridge, Kent C., and Terry E. Robinson. "What Is the Role of Dopamine in Reward: Hedonic Impact, Reward Learning, or Incentive Salience?" *Brain Research Reviews* 28, no. 3 (1998): 309–69.

Berridge, Kent C., and Jay Schulkin. "Palatability Shift of a Salt-Associated Incentive During Sodium Depletion." *Quarterly Journal of Experimental Psychology* 41, no. 2 (1989): 121–38.

Berridge, Virginia, and Griffith Edwards. *Opium and the People: Opiate Use in Nineteenth-Century England*. Free Association Books, 1981.

Berridge, Virginia, and Sarah Mars. "History of Addictions." *Journal of Epidemiology & Community Health* 58, no. 9 (2004): 747–50.

Bertino, Mary, Michael L. Abelson, Sandra H. Marglin et al. "A Small Dose of Morphine Increases Intake of and Preference for Isotonic Saline Among Rats." *Pharmacology Biochemistry and Behavior* 29, no. 3 (1988): 617–23. https://doi.org/10.1016/0091-3057(88)90029-9.

Binder, Elisabeth B., and Charles B. Nemeroff. "The CRF System, Stress, Depression and Anxiety—Insights from Human Genetic Studies." *Molecular Psychiatry* 15, no. 6 (2010): 574–88.

Birnie, Matthew T., Cassandra L. Kooiker, Annabel K. Short et al. "Plasticity of the Reward Circuitry After Early-Life Adversity: Mechanisms and Significance." *Biological Psychiatry* 87, no. 10 (2020): 875–84. https://doi.org /10.1016/J.BIOPSYCH.2019.12.018.

Black, Winfried Sweet. *Dope: The Story of the Living Dead*. Star Company, 1928.

Bliss, Michael. "Resurrections in Toronto: Fact and Myth in the Discovery of Insulin." *Bulletin of the American Academy of Arts and Sciences* 38, no. 3 (1984): 15–36.

Bliss, Michael. *William Osler: A Life in Medicine.* Oxford University Press, 1999.

Bodnar, Richard J. "Endogenous Opiates and Behavior." *Peptides* 75 (2016): 18–70.

Bodnar, Richard J., Michael J. Glass, and James E. Koch. "Analysis of Central Opioid Receptor Subtype Antagonism of Hypotonic and Hypertonic Saline Intake in Water-Deprived Rats." *Brain Research Bulletin* 36, no. 3 (1995): 293–300.

Boecker, Henning, Till Sprenger, Mary E. Spilker et al. "The Runner's High: Opioidergic Mechanisms in the Human Brain." *Cerebral Cortex* 18, no. 11 (2008): 2523–31.

Boehm, Christopher. *Moral Origins: The Evolution of Virtue, Altruism, and Shame.* Soft Skull, 2012.

Booth, Martin. *Opium: A History.* Macmillan, 1999.

Brent, Joseph. *Charles Sanders Peirce: A Life.* Indiana University Press, 1998.

Bruchas, Michael R., Benjamin B. Land, and Charles Chavkin. "The Dynorphin/Kappa Opioid System as a Modulator of Stress-Induced and Pro-Addictive Behaviors." *Brain Research* 1314 (2010): 44–55.

Bruchas, Michael R., Benjamin B. Land, Julia C. Lemos, and Charles Chavkin. "CRF1-R Activation of the Dynorphin/Kappa Opioid System in the Mouse Basolateral Amygdala Mediates Anxiety-Like Behavior." *PloS One* 4, no. 12 (2009): e8528. https://doi.org/10.1371/journal.pone.0008528.

Bruijnzeel, Adrie W. "Kappa-Opioid Receptor Signaling and Brain Reward Function." *Brain Research Reviews* 62, no. 1 (2009): 127–46.

Burnham, John C. *Bad Habits: Drinking, Smoking, Taking Drugs, Gambling, Sexual Misbehavior, and Swearing in American History.* New York University Press, 1993.

Cadet, Jean Lud, Christie Brannock, Bruce Ladenheim et al. "Enhanced Upregulation of CRH mRNA Expression in the Nucleus Accumbens of Male Rats After a Second Injection of Methamphetamine Given Thirty Days Later." *PloS One* 9, no. 1 (2014): e84665. https://doi.org/10.1371/JOURNAL.PONE.0084665.

Cajal, Santiago Ramon Y. *Recollections of My Life,* vol. 8. MIT Press, 1989.

Calkins, Alonzo. *Opium and the Opium-Appetite: With Notices of Alcoholic Beverages, Cannabis Indica, Tobacco and Coca, and Tea and Coffee, in Their Hygienic Aspects and Pathologic Relationships.* Lippincott, 1871.

Campbell, Nancy D. "The Construction of Pregnant Drug-Using Women as Criminal Perpetrators." *Fordham Urban Law Journal* 33 (2005): 463.

Campbell, Nancy D. "The Impact of Changes in Neuroscience and Research Ethics on the Intellectual History of Addiction Research." In *Addiction Neuroethics*, 197–213. Academic Press, 2012.

Campbell, Nancy D. *OD: Naloxone and the Politics of Overdose.* MIT Press, 2020.

Campbell, Nancy D. "Toward a Critical Neuroscience of 'Addiction.'" *BioSocieties* 5, no. 1 (2010): 89–104.

Campbell, Nancy D. "When Should Screening and Surveillance Be Used During Pregnancy?" *AMA Journal of Ethics* 20, no. 3 (2018): 288–95.

Campbell, Nancy D., and Elizabeth Ettorre. *Gendering Addiction: The Politics of Drug Treatment in a Neurochemical World.* Springer, 2011.

Campbell, Nancy D., and Anne M. Lovell. "The History of the Development of Buprenorphine as an Addiction Therapeutic." *Annals of the New York Academy of Sciences* 1248, no. 1 (2012): 124–39.

Campbell, Nancy D., J. P. Olsen, and Luke Walden. *The Narcotic Farm: The Rise and Fall of America's First Prison for Drug Addicts.* University Press of Kentucky, 2021.

Cannon, Walter Bradford. *The Wisdom of the Body.* Norton, 1939.

Carmack, Stephanie A., Robin J. Keeley, Janaina C. M. Vendruscolo et al. "Heroin Addiction Engages Negative Emotional Learning Brain Circuits in Rats." *Journal of Clinical Investigation* 129, no. 6 (2019): 2480–84.

Carmack, Stephanie A., Janaina C. M. Vendruscolo, M. Adrienne McGinn et al. "Corticosteroid Sensitization Drives Opioid Addiction." *Molecular Psychiatry* 27, no. 5 (2022): 2492–2501. https://doi.org/10.1038/s41380-022-01501-1.

Carroll, Marilyn E., and John R. Smethells. "Sex Differences in Behavioral Dyscontrol: Role in Drug Addiction and Novel Treatments." *Frontiers in Psychiatry* 6 (2016): 175.

Castro, Daniel C., and Kent C. Berridge. "Opioid Hedonic Hotspot in Nucleus Accumbens Shell: Mu, Delta, and Kappa Maps for Enhancement of Sweetness 'Liking' and 'Wanting.'" *Journal of Neuroscience* 34, no. 12 (2014): 4239–50.

Chaijale, Nayla N., Andre L. Curtis, Susan K. Wood et al. "Social Stress Engages Opioid Regulation of Locus Coeruleus Norepinephrine Neurons and Induces a State of Cellular and Physical Opiate Dependence." *Neuropsychopharmacology* 38, no. 10 (2013): 1833–43. https://doi.org/10.1038/NPP.2013.117.

Chang, Andrew K., Polly E. Bijur, David Esses et al. "Effect of a Single Dose of Oral Opioid and Nonopioid Analgesics on Acute Extremity Pain in the Emergency Department: A Randomized Clinical Trial." *Jama* 318, no. 17 (2017): 1661–67.

Chavkin, Charles. "Dynorphin–Still an Extraordinarily Potent Opioid Peptide." *Molecular Pharmacology* 83, no. 4 (2013): 729–36.

Chavkin, Charles, and George F. Koob. "Dynorphin, Dysphoria, and Dependence: The Stress of Addiction." *Neuropsychopharmacology* 41, no. 1 (2016): 373.

Chesnut, Mary Boykin, Comer Vann Woodward, and Elisabeth Muhlenfeld. *The Private Mary Chesnut: The Unpublished Civil War Diaries*, vol. 773. Oxford University Press, 1984.

Chittka, Lars. *The Mind of a Bee*. Princeton University Press, 2022.

Cicero, Theodore J., Matthew S. Ellis, Hilary L. Surratt, and Steven P. Kurtz. "The Changing Face of Heroin Use in the United States: A Retrospective Analysis of the Past 50 Years." *JAMA Psychiatry* 71, no. 7 (2014): 821–26.

Clark, Andy. *Surfing Uncertainty: Prediction, Action, and the Embodied Mind*. Oxford University Press, 2015.

Clark, Claire D. *The Recovery Revolution: The Battle over Addiction Treatment in the United States*. Columbia University Press, 2017.

Clarke, Sharon, and Linda A. Parker. "Morphine-Induced Modification of Quinine Palatability: Effects of Multiple Morphine-Quinine Trials." *Pharmacology Biochemistry and Behavior* 51, nos. 2–3 (1995): 505–8.

Cohen, Zachary D., and Robert J. DeRubeis. "Treatment Selection in Depression." *Annual Review of Clinical Psychology* 14, no. 1 (2018): 209–36. https://doi.org/10.1146/ANNUREV-CLINPSY-050817-084746.

Colloca, Luana, Regine Klinger, Herta Flor, and Ulrike Bingel. "Placebo Analgesia: Psychological and Neurobiological Mechanisms." *Pain* 154, no. 4 (2013): 511–14. https://doi.org/10.1016/J.PAIN.2013.02.002.

Cook, Christian J. "Glucocorticoid Feedback Increases the Sensitivity of the Limbic System to Stress." *Physiology & Behavior* 75, no. 4 (2002): 455–64.

Corder, Gregory, Daniel C. Castro, Michael R. Bruchas, and Grégory Scherrer. "Endogenous and Exogenous Opioids in Pain." *Annual Review of Neuroscience* 41, no. 1 (2018): 453–73.

Corlett, Philip R., Aprajita Mohanty, and Angus W. MacDonald III. "What We Think About When We Think About Predictive Processing." *Journal of Abnormal Psychology* 129, no. 6 (2020): 529.

Costigan, Michael, Joachim Scholz, and Clifford J. Woolf. "Neuropathic Pain: A Maladaptive Response of the Nervous System to Damage." *Annual Review of Neuroscience* 32, no. 1 (2009): 1–32. https://doi.org/10.1146/annurev.neuro.051508.135531.

Courtwright, David T. *The Age of Addiction: How Bad Habits Became Big Business.* Harvard University Press, 2019.

Courtwright, David T. *Dark Paradise: A History of Opiate Addiction in America.* Harvard University Press, 1982.

Courtwright, David T. *Forces of Habit: Drugs and the Making of the Modern World.* Harvard University Press, 2001.

Courtwright, David T. "The Hidden Epidemic: Opiate Addiction and Cocaine Use in the South, 1860–1920." *Journal of Southern History* 49, no. 1 (1983): 57–72.

Courtwright, David T., Herman Joseph, and Don Des Jarlais. *Addicts Who Survived: An Oral History of Narcotic Use in America Before 1965.* University of Tennessee Press, 1989.

Craig, Wallace. "Appetites and Aversions as Constituents of Instincts." *Biological Bulletin* 34, no. 2 (1918): 91–107. https://doi.org/10.2307/1536346.

Crist, Richard C., Benjamin C. Reiner, and Wade H. Berrettini. "A Review of Opioid Addiction Genetics." *Current Opinion in Psychology* 27 (2019): 31–35. https://doi.org/10.1016/J.COPSYC.2018.07.014.

Crockett, Molly J., Barbara R. Braams, Luke Clark et al. "Restricting Temptations: Neural Mechanisms of Precommitment." *Neuron* 79, no. 2 (2013): 391–401.

Crothers, Thomas Davison. *Morphinism and Narcomanias from Other Drugs: Their Etiology, Treatment, and Medicolegal Relations.* W. B. Saunders, 1902.

Crowley, Nicole A., Daniel W. Bloodgood, J. Andrew Hardaway et al. "Dynorphin Controls the Gain of an Amygdalar Anxiety Circuit." *Cell Reports* 14, no. 12 (2016): 2774–83. https://doi.org/10.1016/J.CELREP.2016.02.069.

Cummins, Denise Dellarosa, and Robert Cummins. "Biological Prepared-ness and Evolutionary Explanation." *Cognition* 73, no. 3 (1999): B37–B53.

Dana, Jason, and George Loewenstein. "A Social Science Perspective on Gifts to Physicians from Industry." *Jama* 290, no. 2 (2003): 252–55.

Daniels, Derek, and Jay Schulkin. "Water and Salt Intake in Vertebrates: Endocrine and Behavioral Regulation." In *Encyclopaedia of Animal Behavior*, 569–79. Academic Press, 2010.

Darwin, Charles. *The Expression of the Emotions in Man and Animals*. Oxford University Press, 1872/1998.

Darwin, Charles. *Incectivorous Plants*. John Murray, 1875.

Darwin, Charles. *The Life and Letters of Charles Darwin: Including an Auto-biographical Chapter*, vol. 1. D. Appleton, 1887.

Darwin, Charles. *The Origin of Species*. Mentor Books, 1859/1972.

Darwin, Erasmus. *The Botanic Garden*. T. & J. Swords, 1898.

Davis, Michael. "Are Different Parts of the Extended Amygdala Involved in Fear Versus Anxiety?" *Biological Psychiatry* 44, no. 12 (1998): 1239–47.

Davis, Michael, David L. Walker, Leigh Miles, and Christian Grillon. "Phasic vs Sustained Fear in Rats and Humans: Role of the Extended Amygdala in Fear vs Anxiety." *Neuropsychopharmacology* 35, no. 1 (2010): 105–35.

de Guglielmo, Giordano, Marsida Kallupi, Sharona Sedighim et al. "Dopamine D3 Receptor Antagonism Reverses the Escalation of Oxycodone Self-Administration and Decreases Withdrawal-Induced Hyperalgesia and Irritability-Like Behavior in Oxycodone-Dependent Heterogeneous Stock Rats." *Frontiers in Behavioral Neuroscience* 13 (2020): 292. https://doi.org/10.3389/fnbeh.2019.00292.

De Luca Jr., Laurival A. "A Critique on the Theory of Homeostasis." *Physiology & Behavior* 247 (2022): 113712.

Dennett, Daniel C. *Freedom Evolves*. Penguin UK, 2004.

Denton, Derek. *The Hunger for Salt*. Springer Verlag, 1983.

De Oliveira, Lisandra Brandino, Laurival A. De Luca Jr., and José Vanderlei Menani. "Opioid Activation in the Lateral Parabrachial Nucleus Induces Hypertonic Sodium Intake." *Neuroscience* 155, no. 2 (2008): 350–58.

de Quincey, Thomas. *Confessions of an English Opium Eater*. Penguin, 1821/1994.

De Roode, Jacobus C., Thierry Lefèvre, and Mark D. Hunter. "Self-Medication in Animals." *Science* 340, no. 6129 (2013): 150–51.

Dethier, Vincent Gaston. *The Hungry Fly*. Harvard University Press, 1976.

DeWeerdt, Sarah. "Tracing the US Opioid Crisis to Its Roots." *Nature* 573, no. 7773 (2019): S10–S12.

Dewey, John. *Experience and Nature.* Dover, 1925/1989.

Dewey, John. "The Reflex Arc Concept in Psychology." *Psychological Review* 3, no. 4 (1896): 357.

Diamond, Adele. "Consequences of Variations in Genes That Affect Dopamine in Prefrontal Cortex." *Cerebral Cortex* 17, no. suppl_1 (2007): i161–i170.

Dickinson, Anthony, and Bernard Balleine. "Motivational Control of Goal-Directed Action." *Animal Learning & Behavior* 22, no. 1 (1994): 1–18.

Dill, Brendan, and Richard Holton. "The Addict in Us All." *Frontiers in Psychiatry* 5 (2014): 139. https://doi.org/10.3389/fpsyt.2014.00139.

Dingemanse, Mark, Andreas Liesenfeld, Marlou Rasenberg et al. "Beyond Single-Mindedness: A Figure-Ground Reversal for the Cognitive Sciences." *Cognitive Science* 47, no. 1 (2023): e13230.

Dole, Vincent P. "Biochemistry of Addiction." *Annual Review of Chemistry* 39 (1970): 821–40. 10.1146/annurev.bi.39.070170.004133.

Dole, Vincent P. "Implications of Methadone Maintenance for Theories of Narcotic Addiction." *JAMA* 260, no. 20 (1988): 3025–29.

Dole, Vincent P. and Marie E. Nyswander. "Heroin Addiction—A Metabolic Disease." *Archives of Internal Medicine* 120, no. 1 (1967): 19–24.

Donaldson, Zoe R., and Larry J. Young. "Oxytocin, Vasopressin, and the Neurogenetics of Sociality." *Science* 322, no. 5903 (2008): 900–904.

Donnelly, Ellen A., Jascha Wagner, Tammy L. Anderson, and Daniel O'Connell. "Revisiting Neighborhood Context and Racial Disparities in Drug Arrests Under the Opioid Epidemic." *Race and Justice* 12, no. 2 (2022): 322–43.

Dooley, Joe, Lianne Gerber-Finn, Irwin Antone et al. "Buprenorphine-Naloxone Use in Pregnancy for Treatment of Opioid Dependence: Retrospective Cohort Study of 30 Patients." *Canadian Family Physician* 62, no. 4 (2016): e194–e200.

Dormandy, Thomas. *Opium: Reality's Dark Dream.* Yale University Press, 2012.

Dowell, Deborah, Tamara M. Haegerich, and Roger Chou. "CDC Guideline for Prescribing Opioids for Chronic Pain—United States, 2016." *Jama* 315, no. 15 (2016): 1624–45.

Dreborg, Susanne, Görel Sundström, Tomas A. Larsson, and Dan Larhammar. "Evolution of Vertebrate Opioid Receptors." *Proceedings of the National Academy of Sciences* 105, no. 40 (2008): 15487–92.

Ducci, Francesca, and David Goldman. "The Genetic Basis of Addictive Disorders." *Psychiatric Clinics* 35, no. 2 (2012): 495–519.

Dunbar, Robin. *Human Evolution: Our Brains and Behavior.* Oxford University Press, 2016.

Dunbar, Robin, Rebecca Baron, Anna Frangou et al. "Social Laughter Is Correlated with an Elevated Pain Threshold." *Proceedings of the Royal Society B: Biological Sciences* 279, no. 1731 (2012): 1161–67. https://doi.org/10.1098/rspb.2011.1373.

Durkheim, Emile. *Suicide: A Study in Sociology*, trans. J. A. Spaulding and G. Simpson. Free Press, 1897/1951.

Elster, Jon. *Ulysses Unbound.* Cambridge University Press, 2000.

Engel, Andreas K., Karl J. Friston, and Danica Kragic, eds. *The Pragmatic Turn: Toward Action-Oriented Views in Cognitive Science.* Vol. 18. MIT Press, 2016. https://doi.org/10.7551/mitpress/9780262034326.001.0001.

Epstein, Alan N. "Mineralocorticoids and Cerebral Angiotensin May Act Together to Produce Sodium Appetite." *Peptides* 3, no. 3 (1982): 493–94.

Erb, Suzanne, Natalina Salmaso, Demetra Rodaros, and Jane Stewart. "A Role for the CRF-Containing Pathway from Central Nucleus of the Amygdala to Bed Nucleus of the Stria Terminalis in the Stress-Induced Reinstatement of Cocaine Seeking in Rats." *Psychopharmacology* 158 (2001): 360–65. https://doi.org/10.1007/S002130000642.

Erb, Suzanne, and Jane Stewart. "A Role for the Bed Nucleus of the Stria Terminalis, but not the Amygdala, in the Effects of Corticotropin-Releasing Factor on Stress-Induced Reinstatement of Cocaine Seeking." *Journal of Neuroscience* 19, no. 20 (1999): RC35.

Everitt, Barry J., David Belin, Daina Economidou et al. "Neural Mechanisms Underlying the Vulnerability to Develop Compulsive Drug-Seeking Habits and Addiction." *Philosophical Transactions of the Royal Society B: Biological Sciences* 363, no. 1507 (2008): 3125–35. https://doi.org/10.1098/RSTB.2008.0089.

Everitt, Barry J., and Trevor W. Robbins. "Drug Addiction: Updating Actions to Habits to Compulsions Ten Years On." *Annual Review of Psychology* 67, no. 1 (2016): 23–50.

Fillingim, Roger B., Christopher D. King, Margarete C. Ribeiro-Dasilva et al. "Sex, Gender, and Pain: A Review of Recent Clinical and Experimental Findings." *Journal of Pain* 10, no. 5 (2009): 447–85. https://doi.org/10.1016/J.JPAIN.2008.12.001.

Fiorentine, Robert, and Maureen P. Hillhouse. "Drug Treatment and 12-Step Program Participation: The Additive Effects of Integrated Recovery Activities." *Journal of Substance Abuse Treatment* 18, no. 1 (2000): 65–74.

Fisher, Carl Erik. *The Urge: Our History of Addiction*. Penguin, 2022.

Fitzsimons, James T. "Angiotensin, Thirst, and Sodium Appetite." *Physiological Reviews* 78, no. 3 (1998): 583–686.

Fitzsimons, James T. *The Physiology of Thirst and Sodium Appetite*. Monographs of the Physiological Society, no. 35. Oxford University Press, 1979.

Flagel, Shelly B., Sraboni Chaudhury, Maria Waselus et al. "Genetic Background and Epigenetic Modifications in the Core of the Nucleus Accumbens Predict Addiction-Like Behavior in a Rat Model." *Proceedings of the National Academy of Sciences* 113, no. 20 (2016): E2861–E2870. https://doi.org/10.1073/PNAS.1520491113.

Fletcher, Anne M. *Inside Rehab: The Surprising Truth About Addiction Treatment—And How to Get Help That Works*. Penguin, 2013.

Fluharty, Steven J., and Alan N. Epstein. "Sodium Appetite Elicited by Intracerebroventricular Infusion of Angiotensin II in the Rat: II. Synergistic Interaction with Systemic Mineralocorticoids." *Behavioral Neuroscience* 97, no. 5 (1983): 746.

Flynn, Francis W., Harvey J. Grill, Jay Schulkin, and Ralph Norgren. "Central Gustatory Lesions: II. Effects on Sodium Appetite, Taste Aversion Learning, and Feeding Behaviors." *Behavioral Neuroscience* 105, no. 6 (1991): 944. https://doi.org/10.1037/0735-7044.105.6.944.

Foley, Kathleen M. "The Treatment of Cancer Pain." *New England Journal of Medicine* 313, no. 2 (1985): 84–95.

Foote, Jeffrey, Carrie Wilkens, Nicole Kosanke, and Stephanie Higgs. *Beyond Addiction: How Science and Kindness Help People Change*. Simon & Schuster, 2014.

Fort, Joel. *The Pleasure Seekers: The Drug Crisis, Youth, and Society*. Bobbs-Merrill, 1969.

François, Amaury, Sarah A. Low, Elizabeth I. Sypek et al. "A Brainstem-Spinal Cord Inhibitory Circuit for Mechanical Pain Modulation by GABA and Enkephalins." *Neuron* 93, no. 4 (2017): 822–39.

Gallistel, Charles R. *The Organization of Action*. Earlbaum, 1980.

Garcia, John, and Robert A. Koelling. "Relation of Cue to Consequence in Avoidance Learning." *Psychonomic Science* 4 (1966): 123–24.

Gaston, Shytierra. "Enforcing Race: A Neighborhood-Level Explanation of Black–White Differences in Drug Arrests." *Crime & Delinquency* 65, no. 4 (2019): 499–526.

Geneviève, Lester Darryl, Andrea Martani, David Shaw et al. "Structural Racism in Precision Medicine: Leaving No One Behind." *BMC Medical Ethics* 21 (2020): 1–13.

Gigerenzer, Gerd. *Adaptive Thinking*. Oxford University Press, 2000.

Gilbert, Sam J., Peter M. Gollwitzer, Anna-Lisa Cohen et al. "Separable Brain Systems Supporting Cued Versus Self-Initiated Realization of Delayed Intentions." *Journal of Experimental Psychology: Learning, Memory, and Cognition* 35, no. 4 (2009): 905.

Gintzler, Alan R., Michael D. Gershon, and Sydney Spector. "A Nonpeptide Morphine-Like Compound: Immunocytochemical Localization in the Mouse Brain." *Science* 199, no. 4327 (1978): 447–48.

Goethe, Johan W. *The Metamorphosis of Plants*. MIT Press, 1790/2009.

Gold, Philip W. "The Organization of the Stress System and Its Dysregulation in Depressive Illness." *Molecular Psychiatry* 20, no. 1 (2015): 32–47. http://doi.org/10.1038/mp.2014.163.

Goldberg, Jeff. *Anatomy of a Scientific Discovery*. Bantam, 1988.

Goldberg, Simon B., Sin U. Lam, Willoughby B. Britton, and Richard J. Davidson. "Prevalence of Meditation-Related Adverse Effects in a Population-Based Sample in the United States." *Psychotherapy Research* 32, no. 3 (2022): 291–305.

Goldstein, Avram. "Opioid Peptides Endorphins in Pituitary and Brain: Studies on Opiate Receptors Have Led to Identification of Endogenous Peptides with Morphine-Like Actions." *Science* 193, no. 4258 (1976): 1081–86.

Gollwitzer, Peter M. "Implementation Intentions: Strong Effects of Simple Plans." *American Psychologist* 54, no. 7 (1999): 493.

Gollwitzer, Peter M. "Mindset Theory of Action Phases." *Handbook of Theories of Social Psychology* 1 (2012): 526–45.

Goodman, Aviel. "Neurobiology of Addiction: An Integrative Review." *Biochemical Pharmacology* 75, no. 1 (2008): 266–322.

Gopnik, Alison. "Childhood as a Solution to Explore–Exploit Tensions." *Philosophical Transactions of the Royal Society B* 375, no. 1803 (2020): 20190502.

Gould, Stephen Jay. *The Structure of Evolutionary Theory*. Cambridge University Press, 2002.

Gray, J. Megan, Christopher D. Wilson, Tiffany T. Y. Lee et al. "Sustained Glucocorticoid Exposure Recruits Cortico-Limbic CRH Signaling to Modulate Endocannabinoid Function." *Psychoneuroendocrinology* 66 (2016): 151–58.

Graybiel, Ann M. "The Basal Ganglia and Chunking of Action Repertoires." *Neurobiology of Learning and Memory* 70, nos. 1–2 (1998): 119–36.

Graybiel, Ann M., and Kyle S. Smith. "Good Habits, Bad Habits." *Scientific American* 310, no. 6 (2014): 38–43. https://doi.org/10.1038/SCIENTIFIC AMERICAN0614-38.

Greene, Jeremy A., and David Herzberg. "Hidden in Plain Sight: Marketing Prescription Drugs to Consumers in the Twentieth Century." *American Journal of Public Health* 100, no. 5 (2010): 793–803.

Grisel, Judith. *Never Enough: The Neuroscience and Experience of Addiction.* Anchor, 2019.

Gulland, John Masson, and Robert Robinson. "Constitution of Codeine and Thebaine." *Memoirs and Proceedings of the Manchester Literary and Philosophical Society* 69 (1925): 79–86.

Guo, Li, Thilo Winzer, Xiaofei Yang et al. "The Opium Poppy Genome and Morphinan Production." *Science* 362, no. 6412 (2018): 343–47. https://doi .org/10.1126/science.aat4096.

Haerian, Batoul Sadat, and Monir Sadat Haerian. "OPRM1 rs1799971 Polymorphism and Opioid Dependence: Evidence from a Meta-Analysis." *Pharmacogenomics* 14, no. 7 (2013): 813–24.

Halpern, John H., and David Blistein. *Opium: How an Ancient Flower Shaped and Poisoned Our World.* Hachette, 2019.

Hamington, Maurice. *Embodied Care: Jane Addams, Maurice Merleau-Ponty, and Feminist Ethics.* University of Illinois Press, 2004.

Hancock, Dana B., Christina A. Markunas, Laura J. Bierut, and Eric O. Johnson. "Human Genetics of Addiction: New Insights and Future Directions." *Current Psychiatry Reports* 20 (2018): 1–17. https://doi.org/10.1007 /S11920-018-0873-3.

Hari, Johann. *Lost Connections: Why You're Depressed and How to Find Hope.* Bloomsbury, 2019.

Hart, Carl L. *Drug Use for Grown-Ups: Chasing Liberty in the Land of Fear.* Penguin, 2021.

Hart, Carl L. *High Price: A Neuroscientist's Journey of Self-Discovery That Challenges Everything You Know About Drugs and Society.* Harper, 2013.

Harvey, Laura M., Weihua Fan, Miguel Ángel Cano et al. "Psychosocial Intervention Utilization and Substance Abuse Treatment Outcomes in a Multisite Sample of Individuals Who Use Opioids." *Journal of Substance Abuse Treatment* 112 (2020): 68–75. https://doi.org/10.1016/J.JSAT.2020.01.016.

Hawk, Kathryn F., Federico E. Vaca, and Gail D'Onofrio. "Focus: Addiction: Reducing Fatal Opioid Overdose: Prevention, Treatment and Harm Reduction Strategies." *Yale Journal of Biology and Medicine* 88, no. 3 (2015): 235–45.

Hayter, Alethea. *Opium and the Romantic Imagination*. University of California Press, 1996.

Heilig, Markus. *The Thirteenth Step: Addiction in the Age of Brain Science*. Columbia University Press, 2015.

Heilig, Markus, James MacKillop, Diana Martinez et al. "Addiction as a Brain Disease Revised: Why It Still Matters, and the Need for Consilience." *Neuropsychopharmacology* 46, no. 10 (2021): 1715–23.

Herbert, Joe. "Peptides in the Limbic System: Neurochemical Codes for Co-Ordinated Adaptive Responses to Behavioural and Physiological Demand." *Progress in Neurobiology* 41, no. 6 (1993): 723–91.

Herbert, Joe, and Jay Schulkin. "Neurochemical Coding of Adaptive Responses in the Limbic System." In *Hormones, Brain and Behavior*, 659–89. Academic Press, 2002.

Herzberg, David. "Entitled to Addiction? Pharmaceuticals, Race, and America's First Drug War." *Bulletin of the History of Medicine* 91, no. 3 (2017): 586–623.

Herzberg, David. *White Market Drugs: Big Pharma and the Hidden History of Addiction in America*. University of Chicago Press, 2020.

Heyman, Gene M. *Addiction: A Disorder of Choice*. Harvard University Press, 2009.

Hill, Matthew N., Ryan J. McLaughlin, Brenda Bingham et al. "Endogenous Cannabinoid Signaling Is Essential for Stress Adaptation." *Proceedings of the National Academy of Sciences* 107, no. 20 (2010): 9406–11. https://doi.org/10.1073/PNAS.0914661107.

Hinchliffe, Taylor E., and Ying Xia. "Evolutionary Distribution of the δ-Opioid Receptor: From Invertebrates to Humans." *Neural Functions of the Delta-Opioid Receptor* (2015): 67–87.

Hirsch, Jules. "One Thing Leads to Another." *Journal of Clinical Investigation* 114, no. 8 (2004): 1040–43.

Hlavacova, Natasa, Paul D. Wes, Maria Ondrejcakova et al. "Subchronic Treatment with Aldosterone Induces Depression-Like Behaviours and Gene Expression Changes Relevant to Major Depressive Disorder." *International Journal of Neuropsychopharmacology* 15, no. 2 (2012): 247–65. https://doi.org/10.1017/S1461145711000368.

Hoffman, Kelly M., Sophie Trawalter, Jordan R. Axt, and M. Norman Oliver. "Racial Bias in Pain Assessment and Treatment Recommendations, and False Beliefs About Biological Differences Between Blacks and Whites." *Proceedings of the National Academy of Sciences* 113, no. 16 (2016): 4296–4301. https://doi.org/10.1073/pnas.1516047113.

Holton, Richard, and Kent Berridge. "Compulsion and Choice in Addiction." *Addiction and Choice: Rethinking the Relationship*, 153–70. Oxford University Press, 2017.

Hrdy, Sarah Blaffer. *Mothers and Others: The Evolutionary Origins of Mutual Understanding*. Harvard University Press, 2009.

Hu, Yiheng, Ran Zhao, Peng Xu, and Yuannian Jiao. "The Genome of Opium Poppy Reveals Evolutionary History of Morphinan Pathway." *Genomics, Proteomics and Bioinformatics* 16, no. 6 (2018): 460–62. https://doi.org/10.1016/J.GPB.2018.09.002.

Huebner, Bryce. "The Emptiness of Anger." *Journal of Buddhist Philosophy*, no. 3 (2021): 50–67.

Huebner, Bryce. "Picturing, Signifying, and Attending." *Belgrade Philosophical Annual*, no. 31 (2018): 7–40.

Huebner, Bryce. "Planning and Prefigurative Practice." In *The Philosophy of Daniel Dennett*, 295–327. Oxford University Press, 2017.

Huebner, Bryce, and Jay Schulkin. *Biological Cognition*. Cambridge University Press, 2022.

Huffman, Michael A. "Animal Self-Medication and Ethno-Medicine: Exploration and Exploitation of the Medicinal Properties of Plants." *Proceedings of the Nutrition Society* 62, no. 2 (2003): 371–81.

Hughes, John. "Isolation of an Endogenous Compound from the Brain with Pharmacological Properties Similar to Morphine." *Brain Research* 88, no. 2 (1975): 295–308.

Hughes, J., T. W. Smith, H. W. Kosterlitz et al. "Identification of Two Related Pentapeptides from the Brain with Potent Opiate Agonist Activity." *Nature* 258, no. 5536 (1975): 577–79.

Humphreys, Keith. *Circles of Recovery: Self-Help Organizations for Addictions.* Cambridge University Press, 2004.

Humphreys, Keith, Nicolas B. Barreto, Sheila M. Alessi et al. "Impact of 12 Step Mutual Help Groups on Drug Use Disorder Patients Across Six Clinical Trials." *Drug and Alcohol Dependence* 215 (2020): 108213. https://doi.org/10.1016/j.drugalcdep.2020.108213.

James, William. *The Principles of Psychology.* 2 vols. Dover, 1890/1952.

Janzen, Daniel H. "Why Fruits Rot, Seeds Mold, and Meat Spoils." *American Naturalist* 111, no. 980 (1977): 691–713.

Jennings, Elaine M., Bright N. Okine, Michelle Roche, and David P. Finn. "Stress-Induced Hyperalgesia." *Progress in Neurobiology* 121 (2014): 1–18.

Jennings, Oscar. *On the Cure of the Morphine Habit.* Balliere, Tindall and Cox, 1890.

Jeong, Huijeong, Annie Taylor, Joseph R. Floeder et al. "Mesolimbic Dopamine Release Conveys Causal Associations." *Science* 378, no. 6626 (2022): eabq6740.

Johansen, Ayna B., Håvar Brendryen, Farnad J. Darnell, and Dag K. Wennesland. "Practical Support Aids Addiction Recovery: The Positive Identity Model of Change." *BMC Psychiatry* 13 (2013): 1–11. https://doi.org/10.1186/1471-244X-13-201.

Johnson, Mark L., and Jay Schulkin. *Mind in Nature: John Dewey, Cognitive Science, and a Naturalistic Philosophy for Living.* MIT Press, 2023.

Jones, Hendrée E., Sarah H. Heil, Andjela Baewert et al. "Buprenorphine Treatment of Opioid-Dependent Pregnant Women: A Comprehensive Review." *Addiction* 107 (2012): 5–27. https://doi.org/10.1111/j.1360-0443.2012.04035.x.

Jones, Jonathan S. "Opium Slavery." *Journal of the Civil War Era* 10, no. 2 (2020): 185–212.

Kagan, Jerome. *Surprise, Uncertainty, and Mental Structures.* Harvard University Press, 2002.

Kahneman, Daniel. *Thinking, Fast and Slow.* Farrar, Straus and Giroux, 2011.

Kahneman, Daniel, Paul Slovic, and Amos Tversky. *Judgment Under Uncertainty: Heuristics and Biases.* Cambridge University Press, 1982.

Kamienski, Lukasz. *Shooting Up: A Short History of Drugs and War.* Oxford University Press, 2016.

Kane, Harry Hubbell. *The Hypdermic Injection of Morphine.* Chas L. Bermingham, 1880.

Karimi, Sara, Pegah Azizi, Ali Shamsizadeh, and Abbas Haghparast. "Role of Intra-Accumbal Cannabinoid CB1 Receptors in the Potentiation, Acquisition and Expression of Morphine-Induced Conditioned Place Preference." *Behavioural Brain Research* 247 (2013): 125–31.

Keefe, Patrick Radden. *Empire of Pain: The Secret History of the Sackler Dynasty*. Anchor, 2021.

Klein, Colin. *What the Body Commands: The Imperative Theory of Pain*. MIT Press, 2015.

Kolb, Lawrence Coleman. *Drug Addiction: A Medical Problem*. Charles C. Thomas, 1962.

Kolber, Benedict J., Maureen P. Boyle, Lindsay Wieczorek et al. "Transient Early-Life Forebrain Corticotropin-Releasing Hormone Elevation Causes Long-Lasting Anxiogenic and Despair-Like Changes in Mice." *Journal of Neuroscience* 30, no. 7 (2010): 2571–81. https://doi.org/10.1523/JNEUROSCI.4470-09.2010.

Koo, Ja Wook, Michelle S. Mazei-Robison, Quincey LaPlant et al. "Epigenetic Basis of Opiate Suppression of Bdnf Gene Expression in the Ventral Tegmental Area." *Nature Neuroscience* 18, no. 3 (2015): 415–22.

Koob, George F. "The Dark Side of Emotion: The Addiction Perspective." *European Journal of Pharmacology* 753 (2015): 73–87. https://doi.org/10.1016/j.ejphar.2014.11.044.

Koob, George F. "Drug Addiction: Hyperkatifeia/Negative Reinforcement as a Framework for Medications Development." *Pharmacological Reviews* 73, no. 1 (2021): 163–201. https://doi.org/10.1124/pharmrev.120.000083.

Koob, George F., and Floyd E. Bloom. "Cellular and Molecular Mechanisms of Drug Dependence." *Science* 242, no. 4879 (1988): 715–23.

Koob, George, and Mary Jeanne Kreek. "Stress, Dysregulation of Drug Reward Pathways, and the Transition to Drug Dependence." *American Journal of Psychiatry* 164, no. 8 (2007): 1149–59. https://doi.org/10.1176/appi.ajp.2007.05030503.

Koob, George F., and Michel Le Moal. *Neurobiology of Addiction*. Elsevier, 2006.

Koob, George F., and Jay Schulkin. "Addiction and Stress: An Allostatic View." *Neuroscience & Biobehavioral Reviews* 106 (2019): 245–62. https://doi.org/10.1016/j.neubiorev.2018.09.008.

Koob, George F., and Eric P. Zorrilla. "Neurobiological Mechanisms of Addiction: Focus on Corticotropin-Releasing Factor." *Current Opinion in Investigational Drugs* 11, no. 1 (2010): 63.

Kramer, John C. "The Opiates: Two Centuries of Scientific Study." *Journal of Psychedelic Drugs* 12, no. 2 (1980): 89–103. https://doi.org/10.1080/02791072.1980.10471562.

Kreek, Mary Jeanne, Brian Reed, and Eduardo Butelman. "Current Status of Opioid Addiction Treatment and Related Preclinical Research." *Science Advances* 5, no. 10 (2019): eaax9140.

Kreek, Mary Jeanne, Yan Zhou, Eduardo R. Butelman, and Orna Levran. "Opiate and Cocaine Addiction: From Bench to Clinic and Back to the Bench." *Current Opinion in Pharmacology* 9, no. 1 (2009): 74–80. https://doi.org/10.1016/j.coph.2008.12.016.

Krieckhaus, E. E. "'Innate Recognition' Aids Rats in Sodium Regulation." *Journal of Comparative and Physiological Psychology* 73, no. 1 (1970): 117.

Krieckhaus, E. E., and George Wolf. "Acquisition of Sodium by Rats: Interaction of Innate Mechanisms and Latent Learning." *Journal of Comparative and Physiological Psychology* 65, no. 2 (1968): 197.

Krief, Sabrina, Michael A. Huffman, Thierry Sévenet et al. "Bioactive Properties of Plant Species Ingested by Chimpanzees (Pan troglodytes schweinfurthii) in the Kibale National Park, Uganda." *American Journal of Primatology* 68, no. 1 (2006): 51–71. https://doi.org/10.1002/ajp.20206.

Kropotkin, Peter. *Mutual Aid: A Factor of Evolution*. Black Rose, 1904/2021.

Kuhar, Michael. *The Addicted Brain: Why We Abuse Drugs, Alcohol, and Nicotine*. FT Press, 2011.

Kurtz, Linda Farris, and Michael Fisher. "Participation in Community Life by AA and NA Members." *Contemporary Drug Problems* 30, no. 4 (2003): 875–904. https://doi.org/10.1177/009145090303000407.

Laland, Kevin. *Darwin's Unfinished Symphony: How Culture Made the Human Mind*. Princeton University Press, 2017.

Larhammar, Dan, Christina Bergqvist, and Görel Sundström. "Ancestral Vertebrate Complexity of the Opioid System." *Vitamins & Hormones* 97 (2015): 95–122. https://doi.org/10.1016/bs.vh.2014.11.001.

Larhammar, Dan, Susanne Dreborg, Tomas A. Larsson, and Görel Sundström. "Early Duplications of Opioid Receptor and Peptide Genes in Vertebrate Evolution." *Annals of the New York Academy of Sciences* 1163, no. 1 (2009): 451–53. https://doi.org/10.1111/j.1749-6632.2008.03672.x.

Lavallee, Zoey. "Affective Scaffolding in Addiction." *Inquiry* (2023): 1–29.

Lazarus, Bernard. "A Contribution to the Therapeutic Action of Heroin." *Boston Medical and Surgical Journal* 143, no. 24 (1900): 600–602.

LeDoux, Joseph E. *Anxious: Using the Brain to Understand and Treat Fear and Anxiety.* Viking, 2015.

Lembke, Anna. *Drug Dealer, MD: How Doctors were Duped, Patients got Hooked, and Why It's so Hard to Stop.* Johns Hopkins University Press, 2016.

Lemos, Julia C., Matthew J. Wanat, Jeffrey S. Smith et al. "Severe Stress Switches CRF Action in the Nucleus Accumbens from Appetitive to Aversive." *Nature* 490, no. 7420 (2012): 402–6.

Leshner, Alan I. "Addiction Is a Brain Disease, and It Matters." *Science* 278, no. 5335 (1997): 45–47. https://doi.org/10.1126/science.278.5335.45.

Levinstein, Eduard. *Morbid Craving for Morphia: Die Morphiumsucht.* Smith, Elder, 1878.

Levis, Sophia C., Brandon S. Bentzley, Jenny Molet et al. "On the Early Life Origins of Vulnerability to Opioid Addiction." *Molecular Psychiatry* 26, no. 8 (2021): 4409–16. https://doi.org/10.1038/s41380-019-0628-5.

Levran, Orna, Matthew Randesi, John Rotrosen et al. "A 3'UTR SNP rs885863, a Cis-eQTL for the Circadian Gene VIPR2 and lincRNA 689, Is Associated with Opioid Addiction." *PLoS One* 14, no. 11 (2019): e0224399.

Libby, Ronald T. *Treating Doctors as Drug Dealers: The DEA's War on Prescription Painkillers.* Policy Analysis, 2005.

Liedtke, Wolfgang B., Michael J. McKinley, Lesley L. Walker et al. "Relation of Addiction Genes to Hypothalamic Gene Changes Subserving Genesis and Gratification of a Classic Instinct, Sodium Appetite." *Proceedings of the National Academy of Sciences* 108, no. 30 (2011): 12509–14. https://doi.org/10.1073/pnas.1109199108.

Loewenstein, George. *Exotic Preferences: Behavioral Economics and Human Motivation.* Oxford University Press, 2007.

Loewenstein, George. "A Visceral Account of Addiction." In *Getting Hooked: Rationality and Addiction*, ed. Jon Elster and Ole-Jørgen Skog, 237–245. Cambridge University Press, 1999.

Lundy Jr., Robert F., and Ralph Norgren. "Gustatory System." In *The Rat Nervous System*, 890–920. Academic Press, 2004.

Lüthi, Andreas, and Christian Lüscher. "Pathological Circuit Function Underlying Addiction and Anxiety Disorders." *Nature Neuroscience* 17, no. 12 (2014): 1635 43.

Magen, Eran, Carol S. Dweck, and James J. Gross. "The Hidden Zero Effect: Representing a Single Choice as an Extended Sequence Reduces Impulsive Choice." *Psychological Science* 19, no. 7 (2008): 648–49.

Maier, Steven F., Eric P. Wiertelak, and Linda R. Watkins. "Endogenous Pain Facilitory Systems Antianalgesia and Hyperalgesia." *APS Journal* 1, no. 3 (1992): 191–98.

Makino, Shinya, Phillip W. Gold, and Jay Schulkin. "Corticosterone Effects on Corticotropin-Releasing Hormone mRNA in the Central Nucleus of the Amygdala and the Parvocellular Region of the Paraventricular Nucleus of the Hypothalamus." *Brain Research* 640, nos. 1–2 (1994): 105–12.

Manhapra, Ajay, and William C. Becker. "Pain and Addiction: An Integrative Therapeutic Approach." *Medical Clinics* 102, no. 4 (2018): 745–63. https://doi.org/10.1016/j.mcna.2018.02.013.

Manninen, Sandra, Lauri Tuominen, Robin I. Dunbar et al. "Social Laughter Triggers Endogenous Opioid Release in Humans." *Journal of Neuroscience* 37, no. 25 (2017): 6125–31.

Marchant, Nathan J., Valerie S. Densmore, and Peregrine B. Osborne. "Coexpression of Prodynorphin and Corticotrophin-Releasing Hormone in the Rat Central Amygdala: Evidence of Two Distinct Endogenous Opioid Systems in the Lateral Division." *Journal of Comparative Neurology* 504, no. 6 (2007): 702–15.

Marchette, Renata C. N., Adriana Gregory-Flores, Brendan J. Tunstall et al. "κ-Opioid Receptor Antagonism Reverses Heroin Withdrawal-Induced Hyperalgesia in Male and Female Rats." *Neurobiology of Stress* 14 (2021): 100325. https://doi.org/10.1016/j.ynstr.2021.100325.

Markus, Hazel Rose, and Shinobu Kitayama. "Cultures and Selves: A Cycle of Mutual Constitution." *Perspectives on Psychological Science* 5, no. 4 (2010): 420–30.

Marsh, Nina, Abigail A. Marsh, Mary R. Lee, and René Hurlemann. "Oxytocin and the Neurobiology of Prosocial Behavior." *Neuroscientist* 27, no. 6 (2021): 604–19.

Martin, Caitlin E., Mishka Terplan, and Elizabeth E. Krans. "Pain, Opioids, and Pregnancy: Historical Context and Medical Management." *Clinics in Perinatology* 46, no. 4 (2019): 833–47.

Mason, T. H., and W. B. Hamby. "Relief of Morphine Addiction by Prefrontal Lobotomy." *Journal of the American Medical Association* 136, no. 16 (1948): 1039–40.

McEwen, Bruce S. "Allostasis and the Epigenetics of Brain and Body Health over the Life Course: The Brain on Stress." *JAMA Psychiatry* 74, no. 6 (2017): 551–52.

McEwen, Bruce S. "In Pursuit of Resilience: Stress, Epigenetics, and Brain Plasticity." *Annals of the New York Academy of Sciences* 1373, no. 1 (2016): 56–64.

McEwen, Bruce S. "Physiology and Neurobiology of Stress and Adaptation: Central Role of the Brain." *Physiological Reviews* 87, no. 3 (2007): 873–904.

McEwen, Bruce S. "Protective and Damaging Effects of the Mediators of Stress and Adaptation: Allostasis and Allostatic Load." *Allostasis, Homeostasis, and the Costs of Physiological Adaptation* (2004): 65–98. https://doi .org/10.1017/CBO9781316257081.005.

McEwen, Bruce S. "Stress, Adaptation, and Disease: Allostasis and Allostatic Load." *Annals of the New York Academy of Sciences* 840, no. 1 (1998): 33–44.

McEwen, Bruce S., and Eliot Stellar. "Stress and the Individual: Mechanisms Leading to Disease." *Archives of Internal Medicine* 153, no. 18 (1993): 2093–2101.

McGinty, Jacqueline F., Derek van der Kooy, and Floyd E. Bloom. "The Distribution and Morphology of Opioid Peptide Immunoreactive Neurons in the Cerebral Cortex of Rats." *Journal of Neuroscience* 4, no. 4 (1984): 1104–17. https://doi.org/10.1523/JNEUROSCI.04-04-01104.1984.

McRae, Emily. "Emotions and Choice: Lessons from Tsongkhapa." In *Buddhist Perspectives on Free Will: Agentless Agency?*, ed. Rick Repetti, 170–81. Routledge, 2016.

Meghani, Salimah H., and Neha Vapiwala. "Bridging the Critical Divide in Pain Management Guidelines from the CDC, NCCN, and ASCO for Cancer Survivors." *JAMA Oncology* 4, no. 10 (2018): 1323–24.

Mellars, Paul. "Why Did Modern Human Populations Disperse from Africa ca. 60,000 Years Ago? A New Model." *Proceedings of the National Academy of Sciences* 103, no. 25 (2006): 9381–86.

Melzack, Ronald, and Patrick D. Wall. "Pain Mechanisms: A New Theory." *Science* 150, no. 3699 (1965): 971–79. https://doi.org/10.1126/science.150.3699.971.

Merali, Zul, Hymie Anisman, Jonathan S. James et al. "Effects of Corticosterone on Corticotrophin-Releasing Hormone and Gastrin-Releasing Peptide Release in Response to an Aversive Stimulus in Two Regions of the Forebrain (Central Nucleus of the Amygdala and Prefrontal Cortex)." *European Journal of Neuroscience* 28, no. 1 (2008): 165–72. https://doi .org/10.1111/j.1460-9568.2008.06281.x.

Merali, Zul, Judy McIntosh, Pamela Kent et al. "Aversive and Appetitive Events Evoke the Release of Corticotropin-Releasing Hormone and

Bombesin-Like Peptides at the Central Nucleus of the Amygdala." *Journal of Neuroscience* 18, no. 12 (1998): 4758–66. https://doi.org/10.1523 /JNEUROSCI.18-12-04758.1998.

Merlin, Mark David. *On the Trail of the Ancient Opium Poppy.* Associated University Presses, 1984.

Meyers, Karin. "Free Persons, Empty Selves: Freedom and Agency in Light of the Two Truths." In *Free Will, Agency, and Selfhood in Indian Philosophy,* ed. Matthew R. Dasti and Edwin F. Bryant, 41–67. Oxford University Press, 2013.

Milligan, Barry. "Morphine-Addicted Doctors, the English Opium-Eater, and Embattled Medical Authority." *Victorian Literature and Culture* 33, no. 2 (2005): 541–53.

Minozzi, Silvia, Laura Amato, and Marina Davoli. "Development of Dependence Following Treatment with Opioid Analgesics for Pain Relief: A Systematic Review." *Addiction* 108, no. 4 (2013): 688–98.

Mitchell, Marci R., Kent C. Berridge, and Stephen V. Mahler. "Endocannabinoid-Enhanced 'Liking' in Nucleus Accumbens Shell Hedonic Hotspot Requires Endogenous Opioid Signals." *Cannabis and Cannabinoid Research* 3, no. 1 (2018): 166–70. https://doi.org/10.1089/can.2018.0021.

Mithen, Stephen. *The Prehistory of the Mind: The Cognitive Origins of Art and Science.* Thames and Hudson, 1996.

Mogil, Jeffrey S. "Sex Differences in Pain and Pain Inhibition: Multiple Explanations of a Controversial Phenomenon." *Nature Reviews Neuroscience* 13, no. 12 (2012): 859–66. https://doi.org/10.1038/nrn3360.

Mogil, Jeffrey S., and Benjamin Kest. "Sex Differences in Opioid Analgesia: Of Mice and Women." *Pain Forum* 8, no. 1 (1999): 48–50. https://doi .org/10.1016/S1082-3174(99)70045-0.

Mogil, Jeffrey S., Wendy F. Sternberg, Przemyslaw Marek et al. "The Genetics of Pain and Pain Inhibition." *Proceedings of the National Academy of Sciences* 93, no. 7 (1996): 3048–55. https://doi.org/10.1073/pnas.93.7.3048.

Morales, Ileana, and Kent C. Berridge. "'Liking' and 'Wanting' in Eating and Food Reward: Brain Mechanisms and Clinical Implications." *Physiology & Behavior* 227 (2020): 113152.

Morden, Nancy E., Deanna Chyn, Andrew Wood, and Ellen Meara. "Racial Inequality in Prescription Opioid Receipt—Role of Individual Health Systems." *New England Journal of Medicine* 385, no. 4 (2021): 342–51.

Moreno, Jonathan D. *Deciding Together: Bioethics and Moral Consensus.* Oxford University Press, 1995.

Moreno, Jonathan D. *Impromptu Man: J. L. Moreno and the Origins of Psychodrama, Encounter Culture, and the Social Network.* Bellevue Literary Press, 2014.

Moreno, Jonathan D. *Undue Risk: Secret State Experiments on Humans.* Freeman, 2000.

Moreno, Jonathan D., and Jay Schulkin. *The Brain in Context: A Pragmatic Guide to Neuroscience.* Columbia University Press, 2020.

Morrogh-Bernard, H. C., Ivona Foitová, Z. Yeen et al. "Self-Medication by Orang-Utans (Pongo pygmaeus) Using Bioactive Properties of Dracaena Cantleyi." *Scientific Reports* 7, no. 1 (2017): 16653.

Musto, David F. *The American Disease: Origins of Narcotic Control.* Oxford University Press, 1987.

Myers, Brent, and Beverley Greenwood-Van Meerveld. "Divergent Effects of Amygdala Glucocorticoid and Mineralocorticoid Receptors in the Regulation of Visceral and Somatic Pain." *American Journal of Physiology-Gastrointestinal and Liver Physiology* 298, no. 2 (2010): G295–G303. https://doi.org/10.1152/ajpgi.00298.2009.

Na, Elisa S., Michael J. Morris, and Alan Kim Johnson. "Opioid Mechanisms That Mediate the Palatability of and Appetite for Salt in Sodium Replete and Deficient States." *Physiology & Behavior* 106, no. 2 (2012): 164–70. https://doi.org/10.1016/j.physbeh.2012.01.019.

Nemeroff, Charles B., Erik Widerlöv, Garth Bissette et al. "Elevated Concentrations of CSF Corticotropin-Releasing Factor-Like Immunoreactivity in Depressed Patients." *Science* 226, no. 4680 (1984): 1342–44. https://doi.org/10.1126/science.6334362.

Nesse, Randolph M. *Good Reasons for Bad Feelings: Insights from the Frontier of Evolutionary Psychiatry.* Penguin, 2019.

Nesse, Randolph M., and Kent C. Berridge. "Psychoactive Drug Use in Evolutionary Perspective." *Science* 278, no. 5335 (1997): 63–66. https://doi.org/10.1126/science.278.5335.63.

Nesse, Randolph M., and Jay Schulkin. "An Evolutionary Medicine Perspective on Pain and Its Disorders." *Philosophical Transactions of the Royal Society B* 374, no. 1785 (2019): 20190288. https://doi.org/10.1098/rstb.2019.0288.

Netherland, Julie, and Helena Hansen. "White Opioids: Pharmaceutical Race and the War on Drugs That Wasn't." *BioSocieties* 12, no. 2 (2017): 217–38.

Neville, Robert C. *The Cosmology of Freedom*. Yale University Press, 1974.

Nitabach, Michael N., Jay Schulkin, and Alan N. Epstein. "The Medial Amygdala Is Part of a Mineralocorticoid-Sensitive Circuit Controlling NaCl Intake in the Rat." *Behavioural Brain Research* 35, no. 2 (1989): 127–34.

Nummenmaa, Lauri, Lauri Tuominen, Robin Dunbar et al. "Social Touch Modulates Endogenous µ-Opioid System Activity in Humans." *Neuro-Image* 138 (2016): 242–47.

O'Brien, Charles P., A. Thomas McLellan, Anna Rose Childress, and George E. Woody. "Penn/VA Center for Studies of Addiction." *Neuropharmacology* 56 (2009): 44–47. https://doi.org/10.1016/j.neuropharm.2008.06.030.

Olds, James, and Peter Milner. "Positive Reinforcement Produced by Electrical Stimulation of Septal Area and Other Regions of Rat Brain." *Journal of Comparative and Physiological Psychology* 47, no. 6 (1954): 419.

Osler, William. *The Principles and Practice of Medicine*. Appleton, 1892.

Ostrom, Elinor. *Governing the Commons: The Evolution of Institutions for Collective Action*. Cambridge University Press, 1990.

Papaleo, Francesco, Pierre Kitchener, and Angelo Contarino. "Disruption of the CRF/CRF1 Receptor Stress System Exacerbates the Somatic Signs of Opiate Withdrawal." *Neuron* 53, no. 4 (2007): 577–89. https://doi.org/10.1016/j.neuron.2007.01.022.

Park, Paula E., Joel E. Schlosburg, Leandro F. Vendruscolo et al. "Chronic CRF 1 Receptor Blockade Reduces Heroin Intake Escalation and Dependence-Induced Hyperalgesia." *Addiction Biology* 20, no. 2 (2015): 275–84. https://doi.org/10.1111/adb.12120.

Parsons, Loren H., and Yasmin L. Hurd. "Endocannabinoid Signaling in Reward and Addiction." *Nature Reviews Neuroscience* 16, no. 10 (2015): 579–94.

Pasternak, Gavril W., and Ying-Xian Pan. "Mu Opioids and Their Receptors: Evolution of a Concept." *Pharmacological Reviews* 65, no. 4 (2013): 1257–1317. https://doi.org/10.1124/pr.112.007138.

Pasternak, Gavril W., and Solomon H. Snyder. "Opiate Receptor Binding: Enzymatic Treatments That Discriminate Between Agonist and Antagonist Interactions." *Molecular Pharmacology* 11, no. 4 (1975): 478–84.

Pavlov, Ivan P. *Lectures on Conditioned Reflexes*. International Publishing, 1927/1960.

Pavlov, Ivan P. *The Work of the Digestive Glands*. Charles Griffin, 1897/1902.

Peciña, Susana, and Kent C. Berridge. "Dopamine or Opioid Stimulation of Nucleus Accumbens Similarly Amplify Cue-Triggered 'Wanting' for Reward: Entire Core and Medial Shell Mapped as Substrates for PIT Enhancement." *European Journal of Neuroscience* 37, no. 9 (2013): 1529–40. https://doi.org/10.1111/ejn.12174.

Peciña, Susana, and Kent C. Berridge. "Hedonic Hot Spot in Nucleus Accumbens Shell: Where do μ-Opioids Cause Increased Hedonic Impact of Sweetness?" *Journal of Neuroscience* 25, no. 50 (2005): 11777–86.

Peciña, Susana, Jay Schulkin, and Kent C. Berridge. "Nucleus Accumbens Corticotropin-Releasing Factor Increases Cue-Triggered Motivation for Sucrose Reward: Paradoxical Positive Incentive Effects in Stress?" *BMC Biology* 4 (2006): 1–16.

Peciña, Susana, Kyle S. Smith, and Kent C. Berridge. "Hedonic Hot Spots in the Brain." *Neuroscientist* 12, no. 6 (2006): 500–511.

Peirce, Charles Sanders. "The Fixation of Belief." *Popular Science Monthly*, no. 12 (1877): 1–15.

Peirce, Charles Sanders. *Reasoning and the Logic of Things*, ed. Kenneth Lane Ketner and Hillary Putnam. Harvard University Press, 1899/1992.

Pert, Candace B. *Molecules of Emotion: The Science Behind Mind–Body Medicine*. Scribner, 1997.

Pert, Candace B., and Solomon H. Snyder. "Opiate Receptor: Demonstration in Nervous Tissue." *Science* 179, no. 4077 (1973): 1011–14. https://doi.org/10.1126/science.179.4077.1011.

Pessoa, Luiz. "Embracing Integration and Complexity: Placing Emotion Within a Science of Brain and Behaviour." *Cognition and Emotion* 33, no. 1 (2019): 55–60.

Petrovic, Predrag, Eija Kalso, Karl Magnus Petersson, and Martin Ingvar. "Placebo and Opioid Analgesia—Imaging a Shared Neuronal Network." *Science* 295, no. 5560 (2002): 1737–40.

Petrovic, Predrag, Burkhard Pleger, Ben Seymour et al. "Blocking Central Opiate Function Modulates Hedonic Impact and Anterior Cingulate Response to Rewards and Losses." *Journal of Neuroscience* 28, no. 42 (2008): 10509–16.

Petrovich, Gorica D., Andrea P. Scicli, Richard F. Thompson, and Larry W. Swanson. "Associative Fear Conditioning of Enkephalin mRNA Levels in Central Amygdalar Neurons." *Behavioral Neuroscience* 114, no. 4 (2000): 681.

Pezzulo, Giovanni, Thomas Parr, and Karl Friston. "The Evolution of Brain Architectures for Predictive Coding and Active Inference." *Philosophical Transactions of the Royal Society B* 377, no. 1844 (2022): 20200531.

Platt, Stephen R. *Imperial Twilight: The Opium War and the End of China's Last Golden Age.* Knopf, 2018.

Poldrack, Russell. *Hard to Break: Why Our Brains Make Habits Stick.* Princeton University Press, 2021.

Pomrenze, Matthew B., Simone M. Giovanetti, Rajani Maiya et al. "Dissecting the Roles of GABA and Neuropeptides from Rat Central Amygdala CRF Neurons in Anxiety and Fear Learning." *Cell Reports* 29, no. 1 (2019): 13–21. https://doi.org/10.1016/j.celrep.2019.08.083.

Porter, Jane, and Hershel Jick. "Addiction Rare in Patients Treated with Narcotics." *New England Journal of Medicine* 302, no. 2 (1980): 123.

Porter, Roy. *The Greatest Benefit to Mankind: A Medical History of Humanity.* Norton History of Science. Norton, 1997.

Porter, Roy, and Mikulas Teich, eds. *Drugs and Narcotics in History.* Cambridge University Press, 1995.

Power, Michael L., and Jay Schulkin. *The Evolution of Obesity.* Johns Hopkins University Press, 2009.

Powley, Terry L. "The Ventromedial Hypothalamic Syndrome, Satiety, and a Cephalic Phase Hypothesis." *Psychological Review* 84, no. 1 (1977): 89.

Premack, David, and Ann James Premack. "Origins of Human Social Competence." In *The Cognitive Neurosciences*, ed. Michael S. Gazzaniga, 205–18. MIT Press, 1995.

Priddy, Brittany M., Stephanie A. Carmack, Lisa C. Thomas et al. "Sex, Strain, and Estrous Cycle Influences on Alcohol Drinking in Rats." *Pharmacology Biochemistry and Behavior* 152 (2017): 61–67.

Radley, Jason J., Rachel M. Anderson, Caitlin V. Cosme et al. "The Contingency of Cocaine Administration Accounts for Structural and Functional Medial Prefrontal Deficits and Increased Adrenocortical Activation." *Journal of Neuroscience* 35, no. 34 (2015): 11897–910.

Ramachandran, Vilayanur S., and William Hirstein. "The Perception of Phantom Limbs: The DO Hebb Lecture." *Brain: A Journal of Neurology* 121, no. 9 (1998): 1603–30.

Reader, Simon M., and Kevin N. Laland. "Social Intelligence, Innovation, and Enhanced Brain Size in Primates," *Proceedings of the National Academy of Sciences* 99, no. 7 (2002): 4436–41.

Richardson, George B., Taheera N. Blount, and Blair S. Hanson-Cook. "Life History Theory and Recovery from Substance Use Disorder." *Review of General Psychology* 23, no. 2 (2019): 263–74. https://doi.org/10.1037/gpr0000173.

Richardson, Sarah S. *The Maternal Imprint: The Contested Science of Maternal-Fetal Effects*. University of Chicago Press, 2021.

Richerson, Peter J., and Robert Boyd. *Not by Genes Alone: How Culture Transformed Human Evolution*. University of Chicago Press, 2008.

Richerson, Peter J., and Robert Boyd. "Rethinking Paleoanthropology: A World Queerer Than We Supposed." *Evolution of Mind, Brain, and Culture* (2013): 263–302.

Richter, Curt P. "Salt Appetite of Mammals: Its Dependence on Instinct and Metabolism." *L'instinct dans Ie comportement des animaux et de l'homme. Paris* (1956): 577–629.

Richter, Curt P. "Total Self-Regulatory Functions in Animals and Human Beings." *Harvey Lecture Series* 38, no. 63 (1943).

Rieder, Travis N. *In Pain: A Bioethicist's Personal Struggle with Opioids*. Harper Collins, 2019.

Rieder, Travis N. "Pain Medicine During an Opioid Epidemic Needs More Transparency, not Less." *AJOB Neuroscience* 9, no. 3 (2018): 183–206.

Roberts, Charles Stewart. "H. L. Mencken and the Four Doctors: Osler, Halsted, Welch, and Kelly." *Baylor University Medical Center Proceedings* 23, no. 4, 2010: 377–88.

Robinson, Terry E., and Kent C. Berridge. "The Neural Basis of Drug Craving: An Incentive-Sensitization Theory of Addiction." *Brain Research Reviews* 18, no. 3 (1993): 247–91. https://doi.org/10.1016/0165-0173(93)90013-p.

Roitman, Mitchell F., Elisa Na, Gregory Anderson et al. "Induction of a Salt Appetite Alters Dendritic Morphology in Nucleus Accumbens and Sensitizes Rats to Amphetamine." *Journal of Neuroscience* 22, no. 11 (2002): RC225. https://doi.org/10.1523/JNEUROSCI.22-11-j0001.2002.

Rosen, Jeffrey B., and Jay Schulkin. "From Normal Fear to Pathological Anxiety." *Psychological Review* 105, no. 2 (1998): 325.

Rosenkjær, Sophie, Sigrid Juhl Lunde, Irving Kirsch, and Lene Vase. "Expectations: How and When Do They Contribute to Placebo Analgesia?" *Frontiers in Psychiatry* 13 (2022): 817179.

Rosenwasser, Alan M., Jay Schulkin, and Norman T. Adler. "Anticipatory Appetitive Behavior of Adrenalectomized Rats Under Circadian Salt-Access Schedules." *Animal Learning & Behavior* 16, no. 3 (1988): 324–29.

Ross, Don. "Addiction Is Socially Engineered Exploitation of Natural Biological Vulnerability." In *Evaluating the Brain Disease Model of Addiction*, ed. Nick Heather, Matthew Field, Antony C. Moss, and Sally Satel, 359–72. Routledge, 2022.

Rozin, Paul. "The Evolution of Intelligence and Access to the Cognitive Unconscious." In *Progress in Psychobiology and Physiological Psychology*, ed. James M. Sprague and Alan N. Epstein, 245–80. Academic Press, 1976.

Rozin, Paul. "The Socio-Cultural Context of Eating and Food Choice." In *Food Choice, Acceptance and Consumption*, ed. H. L. Meiselman and H. J. H. MacFie, 83–104. Springer, 1996.

Rozin, Paul. "Specific Aversions as a Component of Specific Hungers." *Journal of Comparative and Physiological Psychology* 64, no. 2 (1967): 237.

Rozin, Paul, and James W. Kalat. "Specific Hungers and Poison Avoidance as Adaptive Specializations of Learning." *Psychological Review* 78, no. 6 (1971): 459.

Rozin, Paul, and Jay Schulkin. "Food Selection." In *Neurobiology of Food and Fluid Intake*, ed. Eduard M. Stricker, 297–328. Plenum, 1990. https://doi.org /10.1007/978-1-4613-0577-4_12.

Sabini, John, and Maury Silver. *Moralities of Everyday Life*. Oxford University Press, 1982.

Salavert, Aurélie, Antoine Zazzo, Lucie Martin et al. "Direct Dating Reveals the Early History of Opium Poppy in Western Europe." *Scientific Reports* 10, no. 1 (2020): 20263. https://doi.org/10.1038/s41598-020-76924-3.

Sanford, Christina A., Marta E. Soden, Madison A. Baird et al. "A Central Amygdala CRF Circuit Facilitates Learning About Weak Threats." *Neuron* 93, no. 1 (2017): 164–78. https://doi.org/10.1016/j.neuron.2016 .11.034.

Sanna, Pietro Pablo, Tatsuyoshi Kawamura, Jin Chen et al. "11β-Hydroxysteroid Dehydrogenase Inhibition as a New Potential Therapeutic Target for Alcohol Abuse." *Translational Psychiatry* 6, no. 3 (2016): e760–e760. https:// doi.org/10.1038/tp.2016.13.

Śāntideva. *Bodhicaryāvatāra*, trans. Kate Crosby and Andrew Skilton. Oxford University Press, 2008.

Sapolsky, Robert M. *Behave*. Penguin, 2017.

Sarokin, David, and Jay Schulkin. *The Corporation: Its History and Future*. Cambridge Scholars Publishing, 2020.

Sartre, Jean Paul. *Being and Nothingness*. Philosophical Library, 1956.

Satel, Sally, and Scott Lilienfeld. *Brainwashed: The Seductive Appeal of Mindless Neuroscience*. Basic Books, 2013.

Scarry, Elaine. *The Body in Pain: The Making and Unmaking of the World*. Oxford University Press, 1985.

Schiff, Paul L. "Opium and Its Alkaloids." *American Journal of Pharmaceutical Education* 66, no. 2 (2002): 188–96.

Schmid, Cullen L., Nicole M. Kennedy, Nicolette C. Ross et al. "Bias Factor and Therapeutic Window Correlate to Predict Safer Opioid Analgesics." *Cell* 171, no. 5 (2017): 1165–75. https://doi.org/10.1016/j.cell.2017.10.035.

Schulkin, Jay. *Adaptation and Well-Being: Social Allostasis*. Cambridge University Press, 2011.

Schulkin, Jay. *The CRF Signal*. Oxford University Press, 2017.

Schulkin Jay. *Curt Richter: A Life in the Laboratory*. Johns Hopkins University Press, 2005.

Schulkin, Jay. *Effort: A Behavioral Neuroscience Perspective on the Will*. Erlbaum, 2007.

Schulkin, Jay. *Oliver Wendell Holmes Jr., Pragmatism and Neuroscience*. Palgrave MacMillan, 2019.

Schulkin, Jay. "A Pragmatist Perspective on Brains, Trust, and Choice." *Journal of Speculative Philosophy* 37, no. 1 (2023): 61–80.

Schulkin, Jay. *Rethinking Homeostasis*. MIT Press, 2003.

Schulkin, Jay. *Roots of Social Sensibility and Neural Function*. MIT Press, 2000.

Schulkin, Jay. *Sodium Hunger: The Search for a Salty Taste*. Cambridge University Press, 1991.

Schulkin, Jay, Bruce S. McEwen, and Philip W. Gold. "Allostasis, Amygdala, and Anticipatory Angst." *Neuroscience & Biobehavioral Reviews* 18, no. 3 (1994): 385–96.

Schulkin, Jay, Maria A. Morgan, and Jeffrey B. Rosen. "A Neuroendocrine Mechanism for Sustaining Fear." *Trends in Neurosciences* 28, no. 12 (2005): 629–35. http://doi.org/10.1016/j.tins.2005.09.009.

Schulkin, Jay, and Peter Sterling. "Allostasis: A Brain-Centered, Predictive Mode of Physiological Regulation." *Trends in Neurosciences* 42, no. 10 (2019): 740–52.

Schüll, Natasha Dow. "Addiction by Design: Machine Gambling in Las Vegas." In *Addiction by Design*. Princeton University Press, 2012.

Scull, Andrew, and Jay Schulkin. "Psychobiology, Psychiatry, and Psychoanalysis: The Intersecting Careers of Adolf Meyer, Phyllis Greenacre, and

Curt Richter." *Medical History* 53, no. 1 (2009): 5–36. https://doi.org/10.1017/s002572730000329x.

Seeley, Randy J., Olivier Galaverna, Jay Schulkin et al. "Lesions of the Central Nucleus of the Amygdala II: Effects on Intraoral NaCl Intake." *Behavioural Brain Research* 59, nos. 1–2 (1993): 19–25.

Shaham, Yavin, Douglas Funk, Suzanne Erb et al. "Corticotropin-Releasing Factor, but not Corticosterone, Is Involved in Stress-Induced Relapse to Heroin-Seeking in Rats." *Journal of Neuroscience* 17, no. 7 (1997): 2605–14.

Shakespeare, William. *Othello.* Classic Books Company, 2001.

Sharpe, Melissa J., Chun Yun Chang, Melissa A. Liu et al. "Dopamine Transients Are Sufficient and Necessary for Acquisition of Model-Based Associations." *Nature Neuroscience* 20, no. 5 (2017): 735–42.

Sigerist, Henry E. "Laudanum in the Works of Paracelsus." *Bulletin of the History of Medicine* 9, no. 5 (1941): 530–44.

Silk, Joan B. "The Adaptive Value of Sociality in Mammalian Groups." *Philosophical Transactions of the Royal Society B: Biological Sciences* 362, no. 1480 (2007): 539–59.

Silverman, Gabriel K., George F. Loewenstein, Britta L. Anderson et al. "Failure to Discount for Conflict of Interest When Evaluating Medical Literature: A Randomised Trial of Physicians." *Journal of Medical Ethics* 36, no. 5 (2010): 265–70.

Simon, Eric J., Jacob M. Hiller, and Irit Edelman. "Stereospecific Binding of the Potent Narcotic Analgesic [3H] Etorphine to Rat-Brain Homogenate." *Proceedings of the National Academy of Sciences* 70, no. 7 (1973): 1947–49.

Simon, Herbert A. "The Architecture of Complexity." *Proceedings of the American Philosophical Society* 106, no. 6 (1962): 470–73.

Simon, Herbert A. *Models of Bounded Rationality: Economic Analysis and Public Policy.* MIT Press, 1982.

Skelton, Kelly H., Dana Oren, David A. Gutman et al. "The CRF1 Receptor Antagonist, R121919, Attenuates the Severity of Precipitated Morphine Withdrawal." *European Journal of Pharmacology* 571, no. 1 (2007): 17–24. https://doi.org/10.1016/j.ejphar.2007.05.041.

Smith, Adam. *The Theory of Moral Sentiments.* Liberty Classics, 1759/1982.

Smith, Craig M., and Andrew J. Lawrence. "Salt Appetite, and the Influence of Opioids." *Neurochemical Research* 43, no. 1 (2018): 12–18. https://doi.org/10.1007/s11064-017-2336-3.

Smith, Craig M., Lesley L. Walker, Tanawan Leeboonngam et al. "Endogenous Central Amygdala Mu-Opioid Receptor Signaling Promotes Sodium Appetite in Mice." *Proceedings of the National Academy of Sciences* 113, no. 48 (2016): 13893–98. https://doi.org/10.1073/pnas.1616664113.

Smith, Euclid O. "Evolution, Substance Abuse and Addiction." In *Evolutionary Medicine*, ed. Wenda R. Trevathan, 375–406. Oxford University Press, 1999.

Smith, Kyle S., and Ann M. Graybiel. "Habit Formation." *Dialogues in Clinical Neuroscience* 18, no. 1 (2016): 33–43. https://doi.org/10.31887/DCNS.2016 .18.1/ksmith.

Snyder, Solomon N. *Brainstorming: The Science and Politics of Opiate Research.* Harvard University Press, 1989.

Spector, Alan C. "Linking Gustatory Neurobiology to Behavior in Vertebrates." *Neuroscience & Biobehavioral Reviews* 24, no. 4 (2000): 391–416.

Spencer, Merianne R., Arialdi M. Miniño, and Margaret Warner. "Drug Overdose Deaths in the United States, 2001–2021." *NCHS Data Brief*, no. 457 (2022). https://doi.org/10.15620/cdc:122556.

Spinoza, Benedict de. "Ethics." In *Spinoza: Complete Works*, trans. Samuel Shirley, ed. Michael Morgan, 213–382. Hackett, 1677/2002.

Starling, Ernest Henry. "The Croonian Lectures." *Lancet* 26 (1905): 579–83.

Stefano, George, Radek Ptacek, Hana Kuzelova, and Richard M. Kream. "Endogenous Morphine: Up-to-Date Review 2011." *Folia biologica* 58, no. 2 (2012): 49.

Stein, Christoph. "Opioid Receptors." *Annual Review of Medicine* 67, no. 1 (2016): 433–51.

Stellar, James R., and Eliot Stellar. "Physiological Aspects of Motivation and Reward." *Neurobiology of Motivation and Reward* (1985): 51–82.

Sterelny, Kim. *The Pleistocene Social Contract: Culture and Cooperation in Human Evolution.* Oxford University Press, 2021.

Sterling, Peter. *What Is Health? Allostasis and the Evolution of Human Design.* MIT Press, 2020.

Sterling, Peter, and Joseph Eyer. "Allostasis: A New Paradigm to Explain Arousal Pathology." In *Handbook of Life Stress, Cognition and Health*, ed. Shirley Fisher and James Reason, 629–49. Wiley, 1988.

Sterling, Peter, and Simon Laughlin. *Principles of Neural Design.* MIT Press, 2015.

Strand, Fleur L. *Neuropeptides: Regulators of Physiological Processes.* MIT Press, 1999.

Substance Abuse and Mental Health Services Administration. "Key Substance Use and Mental Health Indicators in the United States: Results from the 2021 National Survey on Drug Use and Health." HHS Publication No. PEP22-07-01-005, NSDUH Series H-57. Center for Behavioral Health Statistics and Quality, Substance Abuse and Mental Health Services Administration, 2022. https://www.samhsa.gov/data/report/2021 -nsduh-annual-national-report.

Swanson, Larry, and Donna. M. Simmons. "Differential Steroid Hormone and Neural Influences on Peptide mRNA Levels in CRH Cells of the Paraventricular Nucleus: A Hybridization Histochemical Study in the Rat." *Journal of Comparative Neurology* 285, no. 4 (1989): 413–35.

Swiergiel, Artur H., Lorey K. Takahashi, William W. Rubin, and Ned H. Kalin. "Antagonism of Corticotropin-Releasing Factor Receptors in the Locus Coeruleus Attenuates Shock-Induced Freezing in Rats." *Brain Research* 587, no. 2 (1992): 263–68.

Sykes, Rebecca Wragg. *Kindred: Neanderthal Life, Love, Death and Art.* Bloomsbury, 2020.

Szalavitz, Maia. *Unbroken Brain: A Revolutionary New Way of Understanding Addiction.* St. Martin's, 2016.

Szalavitz, Maia. *Undoing Drugs: The Untold Story of Harm Reduction and the Future of Addiction.* Hachette, 2021.

Taha, Sharif A., Ebba Norsted, Lillian S. Lee et al. "Endogenous Opioids Encode Relative Taste Preference." *European Journal of Neuroscience* 24, no. 4 (2006): 1220–26. https://doi.org/10.1111/j.1460-9568.2006.04987.x.

Takahashi, Kazuhiro, Toraichi Mouri, Teiji Yamamoto et al. "Corticotropin-Releasing Hormone in the Human Hypothalamus. Free-Floating Immunostaining Method." *Endocrinologia Japonica* 36, no. 2 (1989): 275–80.

Tarjan, Eva, and Derek A. Denton. "Sodium/Water Intake of Rabbits Following Administration of Hormones of Stress." *Brain Research Bulletin* 26, no. 1 (1991): 133–36. https://doi.org/10.1016/0361-9230(91)90197-r.

Tarjan, Eva, Derek A. Denton, and Richard S. Weisinger. "Corticotropin-Releasing Factor Enhances Sodium and Water Intake/Excretion in Rabbits." *Brain Research* 542, no. 2 (1991): 219–24. https://doi.org/10.1016 /0006-8993(91)91570-q.

Tarr, Bronwyn, Jacques Launay, and Robin I. M. Dunbar. "Music and Social Bonding: 'Self-Other' Merging and Neurohormonal Mechanisms." *Frontiers in Psychology* 5 (2014): 1096.

Tenenbaum, Joshua B., Charles Kemp, Thomas L. Griffiths, and Noah D. Goodman. "How to Grow a Mind: Statistics, Structure, and Abstraction." *Science* 331, no. 6022 (2011): 1279–85.

Terenius, Lars. "Stereospecific Interaction Between Narcotic Analgesics and a Synaptic Plasma Membrane Fraction of Rat Cerebral Cortex." *Acta pharmacologica et toxicologica* 32, nos. 3–4 (1973): 317–20.

Terenius, Lars, and Annika Wahlström. "Search for an Endogenous Ligand for the Opiate Receptor." *Acta Physiologica Scandinavica* 94, no. 1 (1975): 74–81.

Terry, Charles E. "The Development and Causes of Opium Addiction as a Social Problem." *Journal of Educational Sociology* 4, no. 6 (1931): 335–46. https://doi.org/10.2307/2961615.

The Beatles. "Happiness Is a Warm Gun." Track 8, side 1. *The Beatles*. EMI, 1968, LP.

The Velvet Underground. "Heroin." Track 1, side 2. *The Velvet Underground and Nico*. Verve Records, 1967, LP.

Thích Nhất Hạnh. *Understanding Our Mind: 50 Verses on Buddhist Psychology*. Parallax, 2001.

Todd, Rebecca M., and Maria G Manaligod. "Implicit Guidance of Attention: The Priority State Space Framework." *Cortex* 102 (2018): 121–38.

Todd, Rebecca M., Vladimir Miskovic, Junichi Chikazoe, and Adam K. Anderson. "Emotional Objectivity: Neural Representations of Emotions and their Interaction with Cognition." *Annual Review of Psychology* 71, no. 1 (2020): 25–48.

Todes, David P. *Pavlov's Physiology Factory: Experiment, Interpretation, Laboratory Enterprise*. Johns Hopkins University Press, 2002.

Tolami, Hedyeh Fazel, Alireza Sharafshah, Laleh Fazel Tolami, and Parvaneh Keshavarz. "Haplotype-Based Association and in Silico Studies of OPRM1 Gene Variants with Susceptibility to Opioid Dependence Among Addicted Iranians Undergoing Methadone Treatment." *Journal of Molecular Neuroscience* 70 (2020): 504–13.

Toth, Mate, Elizabeth I. Flandreau, Jessica Deslauriers et al. "Overexpression of Forebrain CRH During Early Life Increases Trauma Susceptibility in Adulthood." *Neuropsychopharmacology* 41, no. 6 (2016): 1681–90. https://doi.org/10.1038/npp.2015.338.

Towers, Eleanor B., Brendan J. Tunstall, Mandy L. McCracken et al. "Male and Female Mice Develop Escalation of Heroin Intake and Dependence Following Extended Access." *Neuropharmacology* 151 (2019): 189–94.

Tracey, Irene, and Patrick W. Mantyh. "The Cerebral Signature for Pain Perception and Its Modulation." *Neuron* 55, no. 3 (2007): 377–91.

Valenstein, Elliot S. *Brain Control*. Wiley, 1973.

Valentino, Rita J., and Nora D. Volkow. "Untangling the Complexity of Opioid Receptor Function." *Neuropsychopharmacology* 43, no. 13 (2018): 2514–20.

Veiga, Paula. "Opium: Was It Used as a Recreational Drug in Ancient Egypt?" In *Cultural and Linguistic Transition Explored: Proceedings of the ATrA Closing Workshop*, Trieste, May 25–26, 2016, EUT, 3 (2017): 199–215.

Vendruscolo, Leandro F., Estelle Barbier, Joel E. Schlosburg et al. "Corticosteroid-Dependent Plasticity Mediates Compulsive Alcohol Drinking in Rats." *Journal of Neuroscience* 32, no. 22 (2012): 7563–71.

Vickers, Neil. *Coleridge and the Doctors: 1795–1806*. Oxford University Press, 2004.

Volkow, Nora D., George F. Koob, and A. Thomas McLellan. "Neurobiologic Advances from the Brain Disease Model of Addiction." *New England Journal of Medicine* 374, no. 4 (2016): 363–71.

Volkow, Nora D., Michael Michaelides, and Ruben Baler. "The Neuroscience of Drug Reward and Addiction." *Physiological Reviews* 99, no. 4 (2019): 2115–40. https://doi.org/10.1152/physrev.00014.2018.

Von Humboldt, Alexander. *Cosmos: A Sketch of the Physical Description of the Universe*, vol. 1, trans. E. C. Otté. Johns Hopkins University Press, 1848/1997.

Von Humboldt, Alexander, and Aimé Bonpland. *Essay on the Geography of Plants*. University of Chicago Press, 2009.

Walls, Laura Dassow. *The Passage to Cosmos: Alexander von Humboldt and the Shaping of America*. University of Chicago Press, 2019.

Walters, Edgar T., and Amanda C. de Williams. "Evolution of Mechanisms and Behaviour Important for Pain." *Philosophical Transactions of the Royal Society B* 374, no. 1785 (2019): 20190275.

Wang, Fan, Yu Gao, Zhen Han et al. "A Systematic Review and Meta-Analysis of 90 Cohort Studies of Social Isolation, Loneliness and Mortality." *Nature Human Behaviour* 7, no. 8 (2023): 1307–19.

Wang, Qi. *The Autobiographical Self in Time and Culture*. Oxford University Press, 2013.

Wanner, Nicole M., Mathia L. Colwell, and Christopher Faulk. "The Epigenetic Legacy of Illicit Drugs: Developmental Exposures and Late-Life Phenotypes." *Environmental Epigenetics* 5, no. 4 (2019): dvz022.

Waters, R. Parrish, Marion Rivalan, D. A. Bangasser et al. "Evidence for the Role of Corticotropin-Releasing Factor in Major Depressive Disorder." *Neuroscience & Biobehavioral Reviews* 58 (2015): 63–78. https://doi.org/10.1016/j.neubiorev.2015.07.011.

Watkins, Linda R., Eric P. Wiertelak, Lisa E. Goehler et al. "Neurocircuitry of Illness-Induced Hyperalgesia." *Brain Research* 639, no. 2 (1994): 283–99.

Watts, Alan G., and Graciela Sanchez-Watts. "Region-Specific Regulation of Neuropeptide mRNAs in Rat Limbic Forebrain Neurones by Aldosterone and Corticosterone." *Journal of Physiology* 484, no. 3 (1995): 721–36.

Weera, Marcus M., Abigail E. Agoglia, Eliza Douglass et al. "Generation of a CRF1-Cre Transgenic Rat and the Role of Central Amygdala CRF1 Cells in Nociception and Anxiety-Like Behavior." *Elife* 11 (2022): e67822. https://doi.org/10.7554/ELIFE.67822.

Wegner, Daniel M. *The Illusion of Conscious Will.* MIT Press, 2002.

White, William L. *Slaying the Dragon: The History of Addiction Treatment and Recovery in America.* Chestnut Health Systems, 1998.

Whitehead, Alfred North. *The Function of Reason.* Princeton University Press, 1929.

Whiteley, Cecily M. "Depression as a Disorder of Consciousness." *British Journal for the Philosophy of Science* (2021), https://doi.org/10.1086/716838.

Wikler, Abraham. "On the Nature of Addiction and Habituation." *British Journal of Addiction to Alcohol & Other Drugs* 57, no. 2 (1961): 73–79. https://doi.org/10.1111/j.1360-0443.1961.tb05318.x.

Wilson, Frances. *Guilty Thing: The Life of Thomas De Quincey.* Farrar, Straus and Giroux, 2016.

Wilson, Marlene A., and Alexander J. McDonald. "The Amygdalar Opioid System." *Handbook of Behavioral Neuroscience* 26 (2020): 161–212. https://doi.org/10.1016/B978-0-12-815134-1.00008-8.

Wise, Roy A. "Neurobiology of Addiction." *Current Opinion in Neurobiology* 6, no. 2 (1996): 243–51.

Wolf, George. "Innate Mechanisms for Regulation of Sodium Intake." *Olfaction and Taste*, no. 3 (1969): 548–53.

Wood, Susan K., Hayley E. Walker, Rita J. Valentino, and Seema Bhatnagar. "Individual Differences in Reactivity to Social Stress Predict Susceptibility and Resilience to a Depressive Phenotype: Role of Corticotropin-Releasing Factor." *Endocrinology* 151, no. 4 (2010): 1795–1805.

Wood, Wendy, and David T. Neal. "A New Look at Habits and the Habit-Goal Interface." *Psychological Review* 114, no. 4 (2007): 843.

Yan, Junbao, Jinrong Li, Jianqun Yan et al. "Activation of µ-Opioid Receptors in the Central Nucleus of the Amygdala Induces Hypertonic Sodium Intake." *Neuroscience* 233 (2013): 28–43. https://doi.org/10.1016/j.neuroscience.2012.12.026.

Yuferov, Vadim, Orna Levran, Dmitri Proudnikov et al. "Search for Genetic Markers and Functional Variants Involved in the Development of Opiate and Cocaine Addiction and Treatment." *Annals of the New York Academy of Sciences* 1187, no. 1 (2010): 184–207.

Zellner, Debra A., Kent C. Berridge, Harvey J. Grill, and Joseph W. Ternes. "Rats Learn to Like the Taste of Morphine." *Behavioral Neuroscience* 99, no. 2 (1985): 290. https://doi.org/10.1037//0735-7044.99.2.290.

Zhang, Feng, Li-Ping Wang, Martin Brauner et al. "Multimodal Fast Optical Interrogation of Neural Circuitry." *Nature* 446, no. 7136 (2007): 633–39. https://doi.org/10.1038/nature05744.

Zhao, Juan, Xinrui Qi, Shangfeng Gao et al. "Different Stress-Related Gene Expression in Depression and Suicide." *Journal of Psychiatric Research* 68 (2015): 176–85.

Zhou, Hang, Renato Polimanti, Bao-Zhu Yang et al. "Genetic Risk Variants Associated with Comorbid Alcohol Dependence and Major Depression." *JAMA Psychiatry* 74, no. 12 (2017): 1234–41. https://doi.org/10.1001/jamapsychiatry.2017.3275.

Zorn, Jelle, Oussama Abdoun, Sandrine Sonié, and Antoine Lutz. "Cognitive Defusion Is a Core Cognitive Mechanism for the Sensory-Affective Uncoupling of Pain During Mindfulness Meditation." *Psychosomatic Medicine* 83, no. 6 (2021): 566–78.

Zubieta, Jon-Kar, Joshua A. Bueller, Lisa R. Jackson et al. "Placebo Effects Mediated by Endogenous Opioid Activity on µ-Opioid Receptors." *Journal of Neuroscience* 25, no. 34 (2005): 7754–62. https://doi.org/10.1523/JNEUROSCI.0439-05.2005.

INDEX

GPSR Authorized Representative: Easy Access System Europe, Mustamäe tee 50, 10621 Tallinn, Estonia, gpsr.requests@easproject.com

www.ingramcontent.com/pod-product-compliance
Lightning Source LLC
Chambersburg PA
CBHW032121020426
42334CB00016B/1034